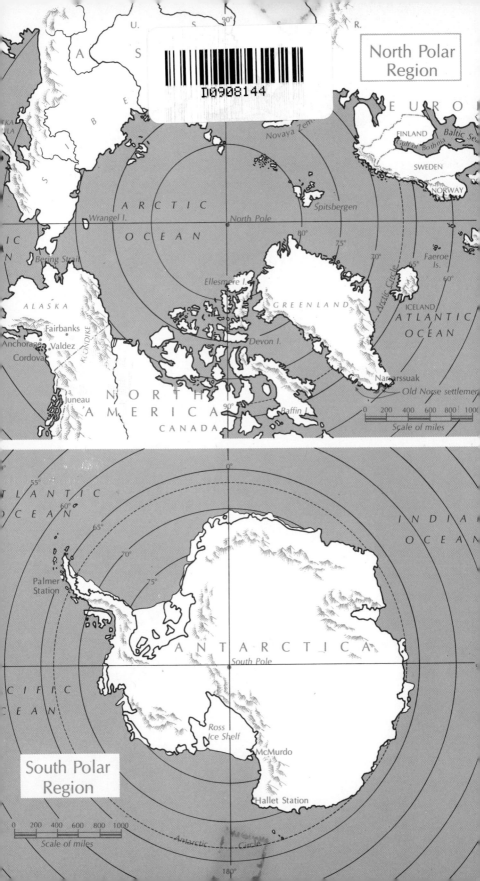

North Polar Region

FINLAND
Gulf of Bothnia
Baltic Sea
SWEDEN
NORWAY

Novaya Zemlya

EURO...

ARCTIC
OCEAN

Spitsbergen

Wrangel I.
North Pole

Bering Strait
80°
75°
70°
65°
60°

Faeroe Is.

Arctic Circle

Ellesmere I.
GREENLAND
ICELAND
ATLANTIC OCEAN

ALASKA
Fairbanks
Anchorage
Valdez
Cordova
KLONDIKE
Juneau

Devon I.

Narssarssuak
Old Norse settlement

NORTH
AMERICA
CANADA
90°
Baffin I.

0 200 400 600 800 1000
Scale of miles

South Polar Region

ATLANTIC OCEAN
55°
60°
65°
70°
75°

INDIAN OCEAN

Palmer Station

ANTARCTICA
South Pole

PACIFIC OCEAN

Ross Ice Shelf
McMurdo
Hallet Station

0 200 400 600 800 1000
Scale of miles

Antarctic Circle
180°

ICE AGE LOST

MAJOR AREAS OF PLEISTOCENE GLACIATION
(SHOWN IN WHITE)

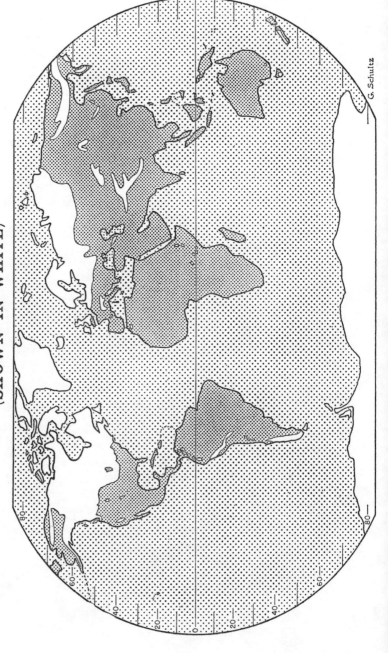

G. Schultz

ICE AGE LOST

Gwen Schultz

1974
ANCHOR PRESS/DOUBLEDAY
Garden City, New York

Frontis map originally published in FRONTIERS magazine (Academy of Natural Sciences of Philadelphia)

PHOTO CREDITS

Daniel K. Eaton, in *Sierra Club Bulletin:* 3
Albert Baumgartner, Munich: 4
National Air Photo Library, Surveys and Mapping Branch,
Department of Energy, Mines and Resources, Canada:
5 (Photo No. T 441-R-208), 6 (Photo No. T 301-L-216),
23 (Photo No. T 127-L-182)
Milwaukee Public Museum: 7, 39
U. S. Navy photograph taken for U. S. Geological Survey: 8, 9
New Zealand National Publicity: 10
Gerald Holdsworth: 11, 12
U. S. Coast Guard: 13
William O. Field, American Geographical Society: 14, 29, 30
U. S. Army Cold Regions Research and Engineering Laboratory: 15
Permission of Z. Burian, Prague: 16, 17, 18
Illinois State Geological Survey: 19
Canadian Government Travel Bureau: 20, 21
Finnish National Tourist Office: 22
Swiss National Tourist Office: 24, 25, 26, 34, 37
Photograph by N. A. Streten, reproduced from the *Journal of
Glaciology* by permission of the International Glaciological
Society: 27
U. S. Geological Survey: 28
Marion T. Millet, American Geographical Society: 31
Official U. S. Navy photograph by Arthur W. Thomas from
U.S.S. *Burton Island:* 32
Norwegian Information Service: 33
H. Peter Wingle: 35, 36
Alaska Travel Division: 38
Anonymity requested: 40

ISBN: 0-385-05759-8
Library of Congress Catalog Card Number: 73–13280
Copyright © 1974 by Gwendolyn M. Schultz
Member of The Authors League of America

TO

MOM AND DAD

WITH

LOVE AND THANKS

PREFACE

ONE DAY in Science Hall library I noticed a student drowsing over a book which was obviously required reading. Finally he folded his arms across it and laid his head down. My curiosity was piqued. What subject was so dull that it put him to sleep? When I had the chance to read the book's title I saw that it was about glaciers. That bothered me, for in my mind glaciers are intensely exciting; but it did not surprise me. For some time I have known that the glacier story was not getting across as well as it might. If university students who have the material handed to them find it hard to acquaint themselves with the subject, how much more difficult it is for others to do so when they do not have access to the specialized literature nor personal contact with field workers and other researchers.

The newest, most informative writing about glaciers and the Ice Age is being done by and for scientists—not in the layman's language and not where the average reader will find it. So despite the public's keen interest in that subject there is a glaring, regrettable lack of available information about it.

Many misconceptions about the Ice Age exist and persist. We see pictures, time and again, of cave men alongside dinosaurs in the Ice Age. A woman living in glaciated farm country proudly tells visitors, "The glacier went right by my house," as though her house had been there at the time and as though the ice as it passed had not been half a mile thick. There are people who will not discuss the marvelous things the ice sheets did in their vicinity because, as they say, "We don't believe in glaciers." There is the all-inclusive way the term "*the* glacier" is nebulously used to mean the doer of everything done by any or all of the glaciers any time and any place in the Ice Age. There is the frequent loose reference to "the polar ice cap" in the North, as though the sea ice were a glacier or Greenland were at the pole. There is the common impression that the whole world turned frigid in the Ice Age. Such misunderstandings, plus questions I have been asked since writing *Glaciers and the Ice Age*, suggested there was a need for a more comprehensive book dealing with these subjects, one that would give an easily understood, composite picture of the Ice Age and bring our knowledge of glaciers up to date.

Therefore, *Ice Age Lost* is written mainly for people who do not visit scientific libraries and have not had recent courses in glacial geology, paleogeography, or anthropology, in order that they may better understand what the Ice Age was all about and what glaciers mean to us. It is hoped that any advanced scholars of the Pleistocene reading this will bear with us through introductory material and elementary explanations where they occur.

This had to be more than a fact book though. One does an injustice to the fabulous Ice Age to write of it in a strictly factual way, and too often glaciers are described and discussed as though they were just cold, sluggish objects, with their thrilling characteristics overlooked. So this book invites you to feel the spell of the Ice Age which still lingers about us and to look at the Ice Age from various points of view.

Besides relating the Ice Age saga I hope to stimulate appreciation of glaciers—their esthetic qualities, their intrinsic value, their long-time role in shaping the world and human development. In pursuing that theme I want to show that glaciers are not merely blank areas as they are shown on maps, that they are not and never have been just idle wastelands where nothing happens, that their influence does not stop at their edge. Along with that aim I want to include as many stirring facts and ideas about glaciers and the Ice Age as can be incorporated in the book's outline and show why it is important that we recognize the significance and worth of our taken-for-granted glaciers.

The world is fast changing and nothing—not even glaciers—can remain undisturbed. Earth science is experiencing a new awakening as upsetting as that of Darwin's time. A few years from now what is said here may seem a bit primitive, as will the contents of most science books, but it will serve to record our stage of understanding at this time and remind us of our weakening link with the Ice Age.

It seems appropriate, almost inevitable, that someone from Wisconsin should write this kind of book, for this state is world-famous glacially. The last massive advance of the ice has been named "The Wisconsin Stage" of glaciation. This state is so richly covered with glacial features, and figures so strongly in the glacial drama, that the Ice Age National Scientific Reserve was established in it. Here the ice's imprint is clear and fresh. Here where I first learned about glaciers, where I have taught students about them, and tramped, ridden, and flown in field trips over the terrain, here in my glacier-made home environment of rounded boulders, ice-smoothed hills, and clustered lakes, I could not help becoming engrossed in the glacier story and inspired by the Ice Age heritage we all share.

My hope is that I can impart to you in a few hours' reading the same thrill of things glacial that over many years I have come to feel.

University of Wisconsin
Madison, Wisconsin
1973

ACKNOWLEDGMENTS

My deep appreciation is expressed to those experts who critically read parts or all of my manuscript and who are responsible for many improvements made in it:

Karl Butzer, Professor of Geography and Anthropology, University of Chicago

Eugene Cameron, Professor of Geology, University of Wisconsin—Madison

John Dallman, Curator of Paleontology and Exhibits, Department of Zoology Museum, University of Wisconsin—Madison

William Drescher, Planning Officer, Department of the Interior, and Research Hydrologist, U. S. Geological Survey

Carl E. Dutton, Geologist, U. S. Geological Survey

Herbert Kubly, writer, traveler, and Professor of English, University of Wisconsin—Parkside

Louis J. Maher, Professor of Geology, University of Wisconsin—Madison

L. G. Monthey, Extension Specialist, Recreation Resources Center, University of Wisconsin—Extension

John T. Robinson, Professor of Zoology, University of Wisconsin—Madison

Wayne Wendland, Meteorology and Geography professor, University of Wisconsin—Madison.

Others to whom I am grateful for assistance are:

Joseph Emielity, Milwaukee Public Museum

Mary Fortney, Map Librarian, Northwestern University

Mary Galneder, Librarian, Map and Air Photo Library, University of Wisconsin—Madison

Leo Johnson, Milwaukee Public Museum

E. J. Moldenhauer, *The Milwaukee Journal*

Randall Sale, Professor of Geography and Cartography, University of Wisconsin—Madison

Walter Scott, conservationist and natural-history historian, Wisconsin Department of Natural Resources

And those persons and agencies that allowed me to use their illustrations, and that helped locate certain hard-to-get photographs.

My thanks also to the many organizations that provided special information, including:

The Admiral Richard E. Byrd Center

Alaska Department of Highways

Alaska Department of Economic Development

Arctic Institute of North America

Arktisk Institut, Denmark

Canadian National Research Council

Canada's Department of Mines and Technical Surveys

Cold Regions Research and Engineering Laboratory, U. S. Army

Denmark's Consulate General

Explorers Club

Forest Service, U. S. Department of Agriculture

Geophysical and Polar Research Center, University of Wisconsin—Madison

National Ice Association

National Park Service, U. S. Department of the Interior

National Science Foundation, Office of Polar Programs

National Weather Service

Norsk Polarinstitutt

Ohio State University, Institute of Polar Studies

The Rand Corporation

Science Service

U. S. Atomic Energy Commission

U. S. Coast Guard

U. S. Geological Survey: Information Office; Water Resources Division; World Data Center A—Glaciology

World Meteorological Organization

State of Washington, Department of Conservation, Division of Power Resources.

CONTENTS

munity activities, construction of buildings and cities,
and the natural environment. Why it is difficult to determine climatic trends.

By contrast, how our lives would be affected if climate
became warmer and drier. Recent heat waves and droughts.

Dangers of climate modification. Experimental cloud
seeding. Plans to redirect ocean currents, warm the Arctic
Ocean, redirect rivers and form lakes, alter the solar-
radiation balance, etc. Effects of air pollution.

The long-range warming trend and retreat of glaciers.
Glacier Bay. Scarcity of snow at Winter Olympics. Cities as
"heat islands." Modern means and incentives to combat
cold.

The increasing need for and scarcity of fresh water.
Glaciers as important reservoirs of this vital resource, holding over three quarters of the world's total supply. Desirable qualities of glacier meltwater. Ambitious plans to
bring water long distances to water-short regions.

The feasibility of towing icebergs to coastal areas needing
water. The special appeal of iceberg ice in beverages. Dealing with icebergs that menace shipping lanes. Icebergs as
aircraft carriers. How icebergs are losing invincibleness
and status.

Glaciers as tourist attractions. Ways in which glaciers are
"eroded" by and for tourists and others. Examples of the

counterpart, tourist "deposition." Tourism in the Alps, Antarctica, Greenland, Alaska, the national parks, the Himalayas, and other glacier regions.

The movement of Early People from the tropics to cold regions. The northward migration of civilization. Recent development of northern North America, Greenland, Siberia, the Arctic Basin, and Antarctica. Acclimatization to cold. Opening frozen shipping routes. Attractions of the cold regions.

"Chopping away" at glaciers. Glacier ice used for refrigeration. Incentives to remove glaciers. The growing mineral shortage, leading to mining under glaciers. Coming methods of melting glaciers and ice cover. Results of sea-level rise if Greenland and Antarctic ice melted.

Indications that living near glaciers has not been detrimental to Man in prehistoric or modern times. Ice Age living conditions—probably more comfortable and pleasant than generally thought. The common desire today to live close to nature, even as Ice Age people did. How our artificial way of life makes winter seem worse than it is. Legacies left to us by the Ice Age.

Our opportunity to see and appreciate what remains of the Ice Age. Disappearance of surviving Ice Age environments and animal life being hastened by the population explosion, our iceward migration, and the scientific age. Our responsibility to preserve our Ice Age legacies.

CHAPTER 1

BORN OF AN ICE AGE

I'll acquaint myself with the glaciers and wild gardens, and get as near the heart of the world as I can.

JOHN MUIR

HOW does an ice age end? As it began—unnoticed. No calendar date marks the return to a springtime world. No formal surrender ceremony proclaims capitulation of glacial forces. There is no trumpet, no tribute, no toast. This change, like most of nature's changes, is gradual and fluctuating.

The appearance of one glacier does not in itself herald the start of an ice age, and likewise an ice age does not prevail until its last vestigial glacier melts to slush and trickles away. No given number of soldiers can by simple definition be called a conquering army, but when the number swells sufficiently and builds up enough combined striking power to advance against all resistance, then a conquering army exists. The ice was such a conqueror. Glaciers marched on every continent. But no army wields control and no world war keeps raging until the last soldier falls. Somewhere along the way the conflict subsides and the war is declared ended. Even so, reserves may be kept ready and the threat of future outbreak may remain.

This book is a salute to our waning Ice Age. It is a study in Ice Age appreciation, a consideration of what the world loses—and gains—when an episode as magnificent as this fades from the scene and from memory.

Was the Ice Age really as awful as generally believed? Where conditions were bad they were terrible. But they weren't bad everywhere. They weren't bad all of the time.

Still, the picture most commonly evoked by the words "Ice Age" is one of specters of ice stalking out of the cold regions. On they came, growing larger, exhaling frigid blasts, sweating muddy rivers, mauling mountains, and smothering the helpless earth with their weighty bodies. All living creatures fled from their path, and anything that remained was devoured by the onward-plodding monsters. No obstacle could halt them. On they pressed, invincible, until they had mutilated nearly one third of the earth's surface.

No wonder we shudder at the thought of the Ice Age. We are relieved to have escaped it; and yet, for some strange reason, we are lured to it. We want to know more about it. We sense that it is close to us, and it is. We feel we are somehow tied to it, and we are.

Today's glaciers are a safe distance away, retiring, resting, or recuperating in the most remote parts of the world—high mountains and polar islands. Although most of them do not affect us directly, we cannot altogether ignore them. We have ambivalent feelings about them. We admire their majestic beauty but fear their sinister strength. In their supine state they seem harmless, but their potential danger and destructive power are never forgotten.

Several times during the Ice Age the glaciers weakened and wasted away, but always they have returned as bold and brash as ever. Although most glaciers are smaller now than in the past and many have disappeared completely, enough are advancing or holding their ground to keep us on guard and remind us that at any time the climatic balance of power could tip in their favor. And what if the glacial army should march again?

Every once in a while that scare becomes real as glaciers here and there surge forward at an unusually rapid rate. Glaciers normally move at a pace too slow to notice, and may even stand still much of the time. But for some unknown reason old quiescent glaciers can suddenly become rejuvenated. They travel so "fast" that one can actually see and hear them move. News media carry little information about their forays because most occur in sparsely populated or uninhabited regions. But back during the "Little Ice Age" of a few centuries ago glaciers on a worldwide front broke loose and advanced into surrounding territory. Even glaciers in populated areas took part in that widespread revival, and anyone living near them was coping directly and helplessly with a new ice age. Later they shrank back again, but not everywhere as far back as they formerly had been. Such concerted glacial strikes are cogent reminders that the ice is ever a menace.

Although glaciers are looked upon as dangerous and bleak things, they attract and intrigue us. Sportsmen, scientists, travelers are drawn to them. The highest mountains have been climbed, the most forbidding seas and wilderness terrains traversed, to reach the aloof and hostile ice. Hardship, loneliness, risk of life, all willingly endured just to be there.

Why do frigid, ostensibly repelling glaciers entice and inspire us? Why is it natural to feel empathy with features as formidable as they? Perhaps this is one answer: We are all born of the Ice Age. Its heritage, in unrealized ways, grips us from within.

Despite our adaptation to an artificial, sophisticated way of life, we came to be what we are while glaciers ruled the earth. Earlier, for millions of mild years before the Ice Age, the primitive progenitors of Man, evolving slowly, were living at animal level, struggling to advance. Then came the Ice Age. During that topsy-turvy time those upward-striving creatures finally found their way to manhood and civilization. At the start of the Ice Age we see them as small, shy, hunted apelike beings confined mainly to warm regions, but before the last ice sheets retreated they had become masters of the animal kingdom and overseers of the whole world.

Many species of animals and Early People became extinct during the Ice Age upheaval. All of the world felt the effects of the glacial transformation in some way, though only a fraction of it was overrun by the ice.

Tropical and polar air masses were in conflict around the globe. Shorelines moved as time after time ocean and lake levels rose and fell. As a result of fluctuating base levels, rivers all over the world were continually reshaping their valleys. Climatic zones and vegetation belts shifted many times. Habitats altered radically. Familiar foods disappeared. New foods took their place. Numerous species of plants and animals died out and others dispersed to different locations. Bands of Early People were obliged or encouraged by circumstances to migrate into new territories, even into neighboring continents, where they met and mixed with other groups. Or, on the other hand, some were prevented from mixing and moving by newly formed barriers of water bodies, landforms, inhospitable territory, or even the ice itself. Those, then, became inbred in isolated societies.

So changes came about, sometimes drastic ones, in a remarkably short evolutionary time span.

The Ice Age world was not static, but rife with conflict and adjustment. We have been brainwashed to think it was entirely unfriendly and austere, but this was a stimulating, challenging time, an evolutionary catalyst which improved the minds, physiques, and skills of Early People, for they were compelled to improvise and adapt at a relatively fast pace if they would escape extinction and maintain an advantage over ever-present rivals. Around the world our Ice Age ancestors walked, commixing into one breed, and winning out over sundry competitors in all climates. We are descendants of those champions. There are glacial genes in us all.

Here are two of the most marvelous dramas of all time—the spreading ice and human evolution—enacted simultaneously on Earth's elaborate revolving stage. The stories are closely intertwined.

The Ice Age shimmers in mystery. Your impression of it is therefore bound to be hazy. Everyone's is. We cannot promise it will be otherwise by book's end. As some of its mysteries are cleared up other unfathomable ones present themselves.

But if one feels inadequately informed about the Ice Age, that is not his fault. The literature is voluminous and many-languaged. Field work rushes on apace around the world. Raw data pile up faster than they can be analyzed. New theories appear continually. Many concepts of only a few years ago have been discarded. Current ones are questioned.

The experts specialize in just limited parts of this complex subject, for in this field as in all others one can explore only so much in one lifetime. For example, one researcher may concentrate mainly on the physics of ice and its movement. Another on Ice Age climatology, or anthropology, or animals, or plants, or stratigraphy, or oceanography, or some other branch; and he or she may narrow that field down eventually to only a certain subtopic or region or period. No one knows the full story. And it is impossible to amalgamate the scattered nuggets of information held in countless minds, libraries, and personal notes. Attempts are made to bring them together through conferences, discussions, field trips, correspondence, and in scholarly literature, but still intercommunication is weak. Besides, it does not (and need not) always lead to agreement. There is wide difference of opinion as to the whats, whys, whens, and wheres of Ice Age events. The sharing and synthesis of information takes time, and for scientists in a hurry there is all too little time.

But here we take a break from the press of data-gathering and debate; sift the facts; and describe in plain language what the current, overall picture of the Ice Age is, at least through one person's eyes. Even fantastic speculations are worth looking into, for some of them prophesy and structure the future.

We want to know what caused the Ice Age in the first place, what the world was like during that time, and how primitive people changed through time into our kind of people. And then there is the commonly asked question about the Ice Age: Will it return? It can be stated right here: no one knows. Another question might be: Has it really gone? Let's look at the facts.

About 30 per cent of the present land area of the world was buried under ice at some time during the Ice Age. Today about 10 per cent still is. Two ice caps of continental proportions grind away vigorously in Greenland and Antarctica. Large healthy icefields are still manufacturing ice in quantity in highlands of North America, Asia, South America, and Europe. Glaciers on a smaller scale are at work too in New Zealand, in Africa, and even on the tropical island of New Guinea. Some exist right along the equator! Besides, about one fifth of the earth's glacier-free land is permafrost, ground that is perennially frozen, in some places over two thousand feet deep. And during winter much more land obediently returns to the ice's clutch, freezing hard; lakes and rivers far from the poles are iced over. Also, 12 per cent of the ocean area is covered with ice part or all of the year. Icebergs roam still more of the ocean and occasionally drift to the subtropics. Snow demonstrates its control over us with staggering force, stopping us in our tracks.

Obviously many aftereffects and remains of the glacial period are still with us, and let us appreciate them for what they are. But the Great Ice Age, when powerful continental ice sheets reached into middle latitudes—that era is gone.

The Ice Age is lost in several ways. It is lost in that the period of maximum glacial dominance is past. The mighty icy climaxes, of which there were several, lie behind us and we are living in a moderately warm period. Perhaps another icy climax lies ahead, but that cannot be foreseen.

The Ice Age is lost, too, in that those glaciers remaining may be doomed to extinction if the present warmth continues and increases. Then our shrinking glaciers are actually fighting for survival, facing weak old age and complete disappearance.

The Ice Age is lost also in that our knowledge of what happened then is meager. We have only squinty peeks into that spectacular time. To have seen the continental glaciers in their prime! Whatever contemporary beings did see them could not comprehend what they saw, nor record their presence there except by their crude artifacts or fragments of their own bones. Tragically the great preponderance of those precious remainders can never be recovered. All but a scattering of strays has already disappeared.

However, the ice itself was not stingy with its souvenirs. What a widespread shambles it left. Wherever it went it dug up the terrain, shoved loose rocks and soil about, gouged bedrock, built gravelly hills, scoured mountains lower, pushed rivers about, and generally remodeled the orderly landscape that regular geologic processes had modeled. With all this glaring evidence of its escapades spread in the open it is almost unbelievable that our educated ancestors of only a few generations ago knew nothing whatever of the Ice Age. The ice had flagrantly left its mark wherever it had been and yet the Ice Age was a totally unknown episode until just the last century. No one before the early 1800's had imagined anything as wild as an Age of Ice. Even now, the concept is so astounding that books arguing that the Ice Age never happened are still being published.

Prehistoric people who had lived with the ice had no way to pass on what they observed. Over a knowledge gap of thousands of years the ice melted away and the weather grew warmer. What did those Early People who saw the edge of a great ice sheet know of its extent? Small glaciers they probably investigated rather thoroughly. The large ones, hardly at all. They must have had the same inquisitiveness we have, so they climbed the sloping edge of the ice, wandered up on it as far as they dared, peered into cracks, shouted to their echoes in crevasses, dropped stones into glassy, smooth-sided holes and listened, tasted the crisp ice and refreshing water, ventured into water-washed tunnels, and picked their way over the slippery, fissured surface. But there was nothing on that endless wasteland for them to pursue. Their feet and legs grew numb as they stood on the ice. Their food, their comforts, their companions were not there. They had no vantage point from which to view the vastness of an ice sheet, but even if they had they would have taken it for granted just as we take the ocean's immensity for granted.

How mountaineers and scientists early in the last century first began to suspect that something as eerie as an Ice Age had occurred in times past is an engrossing story in itself. As geologists came to read the language of the rocks they resurrected that vanished image; but, even while they did, it was fading farther from their ken. The ice continued its retreat, and its handiwork was being continuously erased.

Evidence of the activities of the great gone glaciers is retained in the work they did and in living glaciers which serve as examples of what used to be. But the clues are disappearing rapidly. Chemical and mechanical weathering processes are breaking down the glaciers' sculpturing. Erosion is removing earth materials they deposited, rearranging them, carrying them irretrievably to ocean graves. And the worst eroders of all, people, are at an ever-increasing pace covering up, cluttering, leveling, and demolishing the original landscape. Some of the ice's masterpieces and most informative clues were left in sites that are already dug up or built upon by dense populations. In many areas civilization is erasing the Ice Age record before it can be read.

Part of the priceless record is precariously preserved in surviving glaciers that still contain ice formed long ago. Pressed in the layers of this aged ice are fine silts blown by ancient storms, old meteor dust, ashes of dead volcanoes, pollen of plants that drooped and decayed in primeval obscurity, even the gases of prehistoric atmospheres captured in bubbles. Even as I write, and even as you read, those irreplaceable gems of evidence are being released and lost. Many can still be caught and analyzed to reveal secrets of the past, and some are. But the glaciers keep flowing to their melting edges, spilling out their unrefillable, long-locked treasure chests. When a glacier dies and drains away, nothing of itself remains—no crumbled ruin, no skeleton, no decayed residue. Where once loomed a robust, traveling colossus thousands of feet high is only an empty nothingness of air.

Other Ice Age records are disappearing as the earth's surface is worked over. Rare fossils of extinct animals and of Early People and their artifacts are washed unseen down gullied hillsides and scooped away by power shovels. Scraps of the ledger are still barely readable, but will not be for long, in neatly stratified layers of sediments on old lake bottoms, in terraces along river valleys, in soil profiles developed over the centuries, in pollen and plant distributions, and in surviving animal life.

With most of the Ice Age record forever lost, we should protect and decipher that which is still legible, and prize this remarkable setting in which we are lucky enough to be living.

Our glaciers may be unique in the solar system or universe. Glaciers need special regimes of temperature and precipitation to form and grow. There must be adequate sources of moisture and the right atmospheric circulation to carry that moisture to strategic places and produce precipitation, and low temperatures must preserve it in solid form. What chance is there that on any other planet or moon the atmospheric properties, terrain, and climatic regime would all be ideal as ours have been? Some spheres might be encrusted with ice, but the special conditions needed to produce moving glaciers and ice sheets as we know them may be found nowhere but on our planet. Even here glaciation has been an abnormal occurrence. We live in a truly exceptional period on a most exceptional planet.

What about the white caps on neighbor Mars? Mars, being a tilted, revolving globe like Earth, has its summer and winter seasons as we do. When winter begins in one Martian hemisphere a white spot spreads out over that pole, and as winter continues the little cap expands until it is conspicuous in size. Then with the reversal of seasons it shrinks under the more direct summer sun, while at the opposite pole a similar white cap grows. In this manner an "ice cap" develops at alternate poles at alternate seasons as winter migrates back and forth. Because presumably the amount of moisture on Mars is slight, permitting only thin "ice caps," and because its atmosphere is thin and unprotective compared to ours, the summer sun can easily melt each winter's frozen accumulation.

It seems safe to say Mars does not have such glaciers as we have had. Although that planet is farther from the sun and colder than Earth, it appears to have neither a sufficient moisture supply nor a long enough cold period for large-scale glacier build-up. Some scientists suggest the white Martian caps may not be water ice but rather solid carbon dioxide (dry ice). Others think they may be only snow. Unlike the Martian temporary, thin caps, our ice caps on Greenland and Antarctica are thick, massive accumulations of uncountable years, and how much more imposing were the even grander glaciers at the peaks of our Ice Age!

There used to be faint speculation that glaciers might exist on cloud-masked Venus, but space probes have shown that temperatures there are indeed too high for glaciers.

Earth may well be the glaciers' only home.

Even though the Ice Age lies behind us we are still vitally a part of it. It is natural for us to want to know what our heritage is and what lies ahead—an ice-free world or a resurgence of ice. Although we cannot predict the future, we can better anticipate what climatic direction we are going and plan how to cope with changing conditions, whatever they be, if we reconstruct and understand the past. No drastic climatic change is expected in our lifetime, but change there will be; we shall move toward either greater warmth or greater cold, more snow or less snow, more rain or less rain.

Most of the biography of mankind's prehistoric infancy and adolescence will never be known, but searches into the shadowy past resurrect intriguing fragments of information about those formative years. The clues are incomplete, the story disconnected, but the plot is lively. Prepare to enter that fabulous time.

CHAPTER 2

THE WORLD BEFORE THE ICE ATTACKED

It would seem that man was intended to decipher the past history of his home, for some remnants or traces of all its great events are left as a key to the whole. . . .

I think we may believe that God did not shroud the world he had made in snow and ice without a purpose, and that this, like many other operations of his providence, seemingly destructive and chaotic in its first effects, is nevertheless a work of beneficence and order.

LOUIS AGASSIZ

TO TRAVEL into the past you follow scattered signposts of known facts, indistinct and irregularly placed though they be. Trusty scientists are your guides and interpreters. And the means of transportation is your agile imagination.

Stories are best told from the beginning, so we shall take one long leap back through the years to observe how the Ice Age began and work our way up to the present. First we need to position ourselves in time. Then we shall survey the world before the ice came—its climate and landforms and seas, its plant and animal life. Only by knowing what the pre-glacial world was like can one feel the staggering blow delivered by the Ice Age.

If we start our story before the Ice Age began we are back in the geologic time zone known as the Tertiary Period, so named because it was once considered to be the third main block of time in Earth's history. However, the Tertiary is now judged to have encompassed only 60 to 70 million years, a relatively short block of time in the estimated 5 billion years of Earth's existence. (See Geologic Time Scale.)

GEOLOGIC TIME SCALE

ERA	PERIOD	EPOCH		YEARS AGO (estimate)
Cenozoic	Quaternary	Recent	AGE OF MAN	
		Pleistocene (The Ice Age)		1 to 3 million
	Tertiary	Pliocene	AGE OF MAMMALS	
		Miocene		
		Oligocene		
		Eocene		
		Paleocene		60 to 70 million
Mesozoic	Cretaceous	AGE OF DINOSAURS		
	Jurassic			
	Triassic			230 million
Paleozoic	Permian			
	Carboniferous			
	Devonian			
	Silurian			
	Ordovician			
	Cambrian			600 million
Precambrian	?	Earth formed ?		5 billion

NOTE: There is considerable difference of opinion concerning Pleistocene dates, as explained in the text.

The Tertiary was followed by the Quaternary (accent on the second syllable) Period, which, as its name implies, was considered the fourth block of time, but it is much shorter than even the Tertiary. The terms "Tertiary" and "Quaternary" are falling into disuse among many geologists but acceptable substitutes have yet to be agreed upon.

The Quaternary has two subdivisions—the Pleistocene Epoch followed by the Recent, or Holocene, which includes present time. "Pleistocene" (pronounced *Plice*-tuh-seen) is derived from Greek words meaning "most recent," referring to forms of animal life. "Holocene" means "wholly recent."

The geologic term "The Pleistocene Epoch" has long been considered practically synonymous with "The Ice Age," but now it appears that their respective time spans may be of different lengths unless new definitions are adopted. All geologic boundaries are fuzzy and must be arbitrarily defined, and the Tertiary-Pleistocene boundary is no exception. The Pleistocene Epoch *is* the time of the Ice Age, but lately the Ice Age has been spilling over back into the Tertiary. Its beginnings seem to be earlier than formerly realized.

There are no exact dates for events that long ago, just a relative time scale. The start of the Ice Age used to be set at about a million years ago. That was a neat round number to work from, and so pat that obviously it was not precise. Some scholars would shorten the duration of the Ice Age but most would lengthen it. Two or three million years ago are now favored round-number dates for the start of the Ice Age and the Pleistocene, but many earth scientists would start the Ice Age even earlier, well back in the Tertiary. Later we shall look more closely at the problem of dating Ice Age events, but now it is sufficient to note that there is a broad range of opinion as to when the Ice Age began.

However long the Ice Age was, it was a short part of Earth's life. We might say that the Ice Age got into full swing a couple or a few million years ago; if so, it covered only a wee fraction of 1 per cent of known geologic time, which is estimated to be about 5 billion years. On Earth's long calendar of events the Ice Age was a brief appointment. Not an eon, not an era, not even a period, as geologists reckon time. Just an epoch.

In Tertiary time the world began to look like the world we know. In the deeper past land masses and oceans had varying shapes quite unlike those of today, and it is believed some continents were in different positions; the Atlantic was narrower, so the Americas were closer to Europe and Africa, and Antarctica may have been farther north. The poles seem to have wandered too. But by late Tertiary time continents had the approximate shapes and positions that they have now.

During most of the Tertiary the world differed strikingly from ours in its vegetation cover and the life that inhabited it. As the Tertiary began, between 60 and 70 million years ago, the world was warm and lush, mild everywhere. In fact, during most of Earth's life, climates seem to have been warmer and more uniform over the globe than now, with only moderate cooling toward the poles; seas were broader, covering much of what is now raised land, and glacier ice was apparently uncommon.*

Climate gradually deteriorated in the Tertiary—slowly at first, then more rapidly toward the end, leading into the Ice Age. But during most of the Tertiary, until near its close, all latitudes were warm. There apparently were no steep temperature gradations like those of the present, either from north to south or from summer to winter. Moisture too was more evenly distributed than now. The tropical belt, much broader then, did not just straddle the equator as it does today, but reached in some places more than halfway to the poles. Along the west coast of North America and Europe tropical conditions extended to 50° North Latitude or farther. Beyond the wide, central tropical belt, temperate climate had an expanded range into regions that are now subpolar and polar wastelands. With conditions so benign, worldwide vegetation was luxuriant.

In Antarctica deciduous forests grew where now only algae, mosses, scaly lichens, and scanty grasses squat insecurely. Warm-water marine life lived along its shore. On the other side of the globe forests grew right up to the Arctic Ocean. Greenland then resembled its erroneously flattering name (which was coined by Eric the Red in the tenth century A.D. to entice colonists). In Tertiary time it really was a green land with forests that included such trees as magnolia, lime, redwood, and ginkgo. (The ginkgo tree had wide

* Some ancient glacial-like deposits and erosive work indicate that in the remote past Earth may have experienced large-scale glaciations earlier than the Pleistocene—one at the close of the Precambrian about 700 million years ago, another at the close of the Paleozoic about 200 to 300 million years ago, and perhaps others.

distribution in the Tertiary but nearly became extinct during the Ice Age.) Towering sequoia redwood trees, which now struggle to keep a last roothold in small parts of California and Oregon, flourished all across the Northern Hemisphere before the Ice Age, from western United States and Canada to the Atlantic, even in the Arctic islands, and from England and continental Europe across Siberia to Japan.

In the Tertiary many deciduous trees that we know today were numerous across the Northern Hemisphere, and there was a greater variety of coniferous trees than now.

Warm-climate plants thrived far beyond their present range. Palms were common throughout the middle latitudes. In Europe they grew north into England, France, Germany, and southern Russia, and, in North America, north into Wyoming. Oregon's vegetation resembled present-day rainforests of tropical Central America.

Much of Alaska that is now in permafrost was during pre-glacial time supporting exotic plants like fronded Mexican cycads, breadfruit trees, delicate ferns, magnolias, Oriental water lilies, avocados, low-growing palms, a type of golden rain tree, and ginkgo, cypress, and redwood trees. Also in Alaska were trees of intermediate climates, like willow, poplar, birch, walnut, beech, chestnut, oak, elm, pine, and fir.

Southern England's vegetation before the Ice Age resembled that found now in the Malay Peninsula of Southeast Asia and in Central America. Along her warm rivers crocodiles crawled among palms and basked on banks shaded by fig, breadfruit, and cinnamon trees.

Tropical and subtropical plants were typical of pre-glacial continental Europe. Magnolia, acacia, myrtle, fig, and cinnamon trees, as well as palms, grew in profusion in central Europe along with more familiar temperate trees like oak, beech, and maple and, in cooler parts of the continent, birch and alder. Switzerland enjoyed Mediterranean climate, and lotus flowers bloomed in her then-warm lakes.

A world with climate so congenial and vegetation so extravagant was bound to have a lavish fauna. And so it did. Never before or since has Earth paraded such a decorative array of animal life as it did just prior to the Ice Age. Many animal types would die out during the coming climatic upheaval. Others would adapt to it and change, or migrate.

Little was left of the old reptilian populations which had ruled the world earlier. Gone were the dinosaurs. (They are all too often depicted as living along with Early Man, but they became extinct about 65 million years before he appeared, by the start of the Tertiary.) Only small reptiles such as crocodiles, turtles, snakes, and lizards survived as reminders of the once omnipotent dinosaurian dynasty. Mammals took control in the Tertiary and still retain it. Back in the pre-Tertiary dinosaur days mammals had been inconspicuous little animals, but as the Ice Age approached many grew larger. Some attained tremendous size and took on astonishing forms. Evolution was accelerating, perhaps in response to rapid geographic upsets.

Climatic machinery that would generate the Ice Age was gathering momentum. But animals able to withstand cold had developed sufficiently to take over the world before the onslaught of the ice.

How much better equipped for survival were the warm-blooded, hairy, milk-giving mammals that gave birth to live offspring compared to the cold-blooded, scaly reptiles whose offspring were deposited immaturely in eggs, there to precariously go through their final development before appearing. Reptiles become inactive in low temperature. Being vulnerable to cold and restricted to a warm habitat, they were less mobile than their mammal successors. Mammals—yourself included—have a built-in thermostat. With their body's natural heat-producing and cooling mechanisms they maintain a constant internal temperature and are free to live in a wide range of environments. When the temperature dropped they could adjust to the cold or move, even across a continent or through a whole hemisphere, to reach a more favorable living area. Their hair or fur was insulation against temperature extremes. Their young were protected by being held inside the womb until fully formed. Born highly developed, they were able to travel with the group. The mother's ready milk gave young mammals an added survival and growth advantage.

Mammals developed into such varied forms and became so numerous during the Tertiary that it is called the Golden Age of Mammals. The animal kingdom then was like all the zoos and circuses and wildlife refuges combined. The best observable examples of its grandeur can be seen in decimated groups of its descendants cornered off in Africa and South Asia.

As the Ice Age approached, Earth was convulsing in one of her greatest, most widespread mountain-building periods of all time. It still continues, though not everywhere so spectacularly. Young, rugged mountains were jutting up, and old, rounded ones entered renewed growth periods. The youthful Rockies and Andes, backbones of North and South America, had appeared and were now given new vigorous thrusts upward. Farther west in the United States the Sierra Nevada block was tilting up. Crustal folding and faulting were creating the young Coast Ranges along the Pacific shore. Cracking of the crust allowed lava to pour out onto the surface through fissures, building up broad, thick lava fields in part of that western region; and the Cascade Range, studded with its string of symmetrical volcanic cones, was forming.

Earthquakes, landslides, faulting, and volcanism along that coast attest that tectonic forces are still active there. In fact, the whole Pacific basin rim has been a not very pacific "ring of fire"—a circle of volcanoes and mountain building.

Volcanic activity flared there and in other places throughout the Ice Age. Mountain-making processes of crustal breaking and bending likewise continued. Together they exerted a powerful influence of change on world geography.

The Himalayas, highest of mountains, were reaching toward the top of the world as the Ice Age began. The Alps and other European mountains pushed higher as rock-bending pressure came from the south. Volcanoes showered central France and western Germany with ashes, and floods of lava poured through the forests. The Mediterranean area, the Middle East, and the Rift Valley of East Africa were regions of disturbance. Around the world old mountains were reaching to greater heights and new ones were forcing up their brittle, angular backs, cracking, and sometimes spilling out hot lava. Ocean basins deepened in places and continents stood higher than before.

Erosive rivers gnawed with increased appetites into the newly rising land and spewed copious outpourings of sediments at mountain bases and on ocean floors around the continents.

Not all uplift was violent. There was also gentle rising of coastal plains as along the Atlantic shores of Europe and America. Some land areas that had been separated by the sea were linked, allowing freer movement of animals than in the past. Alaska was joined to Siberia, but that land bridge would be submerged later. Some islands off the coast of Southeast Asia were linked to one another or to the mainland. North and South America were connected as the Ice Age was about to begin, and stayed connected. Europe, Asia, and Africa were joined as they are today. Australia remained isolated.

With these striking geomorphic modifications, climate could not remain unaltered. A geologic revolution was going on and circulation patterns were being rearranged within the atmosphere and oceans, the transporters of heat.

Temperature, precipitation, humidity, air pressure, and winds— the elements that together comprise climate—are acted upon by many controls. The sun is the most powerful one. Others are the distribution of land and water; the trajectories and interplay of air masses and winds; storms and moving centers of high and low pressure; warm and cold ocean currents; variations in altitude.

Glaciers were growing in Antarctica and perhaps in some other isolated mountainous regions even during the Tertiary, before the Ice Age proper began throughout the world.

Certain regions were becoming drier too. Where mountains rose like walls against air masses moving in from the sea they created rain-shadows on their leeward sides. Moist maritime air masses could not everywhere enter the continents' interiors as easily as before. Some first had to rise over mountain heights.

On the way upslope air masses expand, cooling and dropping moisture because cool air cannot hold as much water vapor as warmer air can. Then on the downslope side these air masses produce dry conditions not only because they contain less moisture than before, but also because air by the act of settling becomes more stable and less prone to precipitate, and because compression warms air. As air warms it is capable of holding more moisture and so it evaporates from the surfaces it passes over. This is the case today in the rain-shadows of mountains—the Great Basin and Great Plains of western United States in the Westerly wind belt; the Patagonian Desert in the lee of the Argentine Andes in the Southern Hemisphere Westerlies; the dry steppes and deserts of Inner Asia cut off from the ocean by mountains.

In the expanding arid regions, where rainfall was no longer sufficient to support trees, there was appearing a new kind of vegetation— grass. It would be food for a new kind of animal—the grazing animal. Early forms of horses, cattle, and various other grazers multiplied, utilizing this food. Ultimately grasses would develop into cereal grains which people would learn to cultivate by the end of the Ice

Age. Grain, destined to become their staff of life, would allow them to settle in permanent living sites. But that time was a long way off.

Across the North American scene as the Ice Age was about to begin wandered vast herds of grazing animals. The grasslands abounded with horses and camels and rhinoceroses.

The horse in Early Tertiary had been a small, fox-sized animal with slightly arched back. Gradually it evolved into a larger size and developed into a fast runner, perfectly adapted to the open, unconcealing grasslands. There were ten or more species of the horse in Late Tertiary. During the Ice Age it was to die out in America, while living on in Eurasia. In the sixteenth century Spanish explorers would bring it back to America, where it would thrive again.

Camels, not yet with humps, grazed the North American grasslands in immense herds prior to the glacial epoch and, like the horses, were to abandon America for Eurasia—all except the Andean llama and its relatives. In historic times the camel too was brought back to North America (to the arid West) but unsuccessfully.

As yet there were no cattle, bison, or antelope on the grassy North American steppe, although there was a relative of today's pronghorn "deer." The rhinoceroses were of the hornless as well as the short-legged, hippopotamus-like types. Oreodonts—piglike, cud-chewing animals, now extinct—were present in large numbers.

Carnivores followed the grazing herds and preyed upon them. Most fearsome-looking was the saber-tooth tiger, whose long, curved teeth and exceptionally wide-opening jaw made it a deadly killer. Other cats, wolves, and dog-like attackers were also on the prowl.

On stage in Eurasia as the curtain was about to go up on the Ice Age was a rich assortment of animals. Ancestral elephants and the mastodon were there and the grazers in enormous herds. Rhinoceroses inhabited the grassy plains and bush country. One would have seen cattle, bison, sheep, horses, goat-like animals, bears, civets, tapirs, and many genera of antelope and deer, including the first to sport antlers. Asia had varied giraffes, including one giant form. Dogs, cats, and hyenas attacked the young and stragglers among grazing herbivores. In woodlands and near streams could be found beavers, pigs, porcupines, and hedgehogs.

Waters of the Mediterranean were warm enough to sustain coral growth.

The elephant-like animals that preceded the Ice Age had sundry forms. Some had long jaws, some short jaws. On some the tusks turned up; on some they turned down. Some had long, curved tusks. Some had shovel-like tusks used for scooping vegetation from warm, shallow ponds and lakes. Mastodons and mammoths were to stay on through most of the Ice Age. At the close of the Tertiary, the start of the Ice Age, the kinds of elephants we know appeared, taking their place among the earlier types that remained.

The Tertiary's largest proboscideans (animals with trunks) were the dinotheres, which lived in the warm swamps and marshes of Eurasia and Africa and died out as the Ice Age approached. One fossil found in Romania had a shoulder height of thirteen feet. (An average room is about eight feet high.) Huge as this elephantine beast was, another Tertiary animal outdid him in size—a giant, hornless rhinoceros of central Asia. This, the largest known land mammal of all time, looked like an over-sized, bizarre horse, having a narrow skull and long legs. Preserved parts of one skeleton indicate the live animal was up to twenty-eight feet long and over eighteen feet tall. Largeness in animals, as we shall see, is a trait that carried over into the Pleistocene. But smaller animals need watching too, especially the primates. Some of them will be leading characters in our Ice Age drama.

Among the primates were the monkeys, but they would not become stars. In the Old World they lived beyond their present range, up into central Europe, which was then subtropical. (Today they are found in Europe only at its southwestern tip, Gibraltar.) There were monkeys in the New World too, but no apes, as far as we know. And it is apes we are interested in, for where the manlike apes were there we may possibly find the beginnings of Man.

As the mild Tertiary Period was giving way to the cold Pleistocene, apelike creatures flourished in warm parts of the Old World. Some lived in southern Europe's forests, but a larger number and variety inhabited southern Asia and Africa. Some were following evolutionary paths toward becoming the true apes we know today—gorillas, chimpanzees, orang-utans, and gibbons—while others would evolve along lines leading to the human figure. All others, less fit for survival, were going along paths leading to a dead end. They disappeared.

Certain parts of the world seem to be eliminated as Man's possible "birthplace." Australia appears to be out because primates were not among its fauna and it remained cut off from the larger continents, the developmental areas, during Man's evolution from lower to upper primate form. The Americas also are crossed off the list because, as was mentioned, the highest form of primate known to be living there was the monkey. So we are left with the Old World, especially its tropical or subtropical parts, as the region where Early People probably first appeared. As of now, Africa is considered their most likely place of origin because of the numbers and kinds of fossilized remains and artifacts found there.

Although the high and middle latitudes were feeling the drop in temperature and were donning vegetation more suitable to cold weather, the low latitudes remained warm. Africa, whose middle is crossed by the equator, was not to experience the snow and ice and wintry scenes that higher latitudes would. That continent would have no large glaciers during the Ice Age, only small local ones—in Morocco's High Atlas Mountains and in mountains of Ethiopia, Uganda, Tanzania, and Kenya.

Early People's apelike predecessors very likely lived in, or along open margins of, forests in always-wet-and-warm climatic regions. (In such habitats live the last surviving apes today.) They apparently were primarily vegetarian food-gatherers as apes are now, judging by their tooth structure. In areas that had warmth and rain all year, they found plant foods always available: fruits, nuts, leaves, buds, pods, and berries; and their predominantly vegetarian diets were probably supplemented with birds' eggs, nestlings, insects, and occasionally a small animal. Water was handy; forests by their existence imply ample year-round precipitation. Among the trees would be safety for a nimble climber who had no built-in fighting weapon. Looking at it another way, such weapons of defense and attack as non-tree-dwellers have—hooves, fang teeth, tusks, claws, horns, antlers, tremendous running speed, heavy muscles, thick skin—gave no advantage to a quick leaper among labyrinths of leaves and tangles of branches and vines, and so trends to develop accouterments of that sort were not given preference in the evolutionary selection process. Instead, other traits were favored. A lithe, flexible body. An arm built for locomotion among branches. Versatile hands with jointed fingers and a curvature natural for wrapping around a branch. The loose, opposable thumb, a special boon, could be placed opposite the fingers in various positions to hold things—not just branches—and it was to make all the difference in the world as time went by.

Living among trees helps a creature position itself vertically and look upward. In contrast, strictly terrestrial animals' bodies are horizontal and their attention is directed downward. Tree-dwellers frequently set their body upright—when reaching up for a limb or piece of fruit, when squatting on a branch, when climbing a trunk or sliding down, when brachiating (arm-swinging) from branch to branch. Internal organs in pre-Man became adjusted to that position. Occasionally a front foot, a hand, could be freed and used independently, or both hands could, and so hand-brain coordination was stimulated.

Tree life had endowed Man's precursors with another precious legacy: sharp eyesight with stereoscopic and color vision. In leaps through a network of high branches even slight misjudgment can be fatal. Those individuals who mismaneuvered were dropouts. Perception, foresight, and ability to figure things out gave one an edge over the pressing competition. Those with superior eyesight and other physical assets had an advantage in getting food and fleeing danger. They survived, their genes spread, and their traits edged out less advantageous ones.

While ground-dwellers relied on their sense of smell for detecting danger and tracking other animals, Man's Tertiary tree-dwelling predecessors let that sense play a lesser role, for smell-ability was less useful in a 3D realm of jumps and swings. Their heads, instead of favoring the long snout, gave precedence to a good positioning of the eyes—close together and front-facing—and to brain enlargement. Ground-dwellers who stand on all fours must use their snout not only to investigate but also to pick things up and pull them apart. The working end has to be far enough out front so the eyes can focus on the activity. Sharp teeth disassembled food for eating. Tree-dwellers with obedient fingers did not need that large nose-and-mouth contraption, so what facial protuberance they had inherited gradually diminished as greater reliance was put upon the eyes and hands working together. Killing and butchering were not part of the eating procedure either, at least not yet. The hand picked food, took it apart, and passed it to the mouth. Little need for pointy, ripping teeth or powerful jaws or thick neck muscles.

Even though our forest heritage is obvious, the humid forest regions where pre-Man might well have developed have not given much tangible archeological evidence of his presence. Most early finds are in drier regions. But that need not mean that Early People lived more in the dry regions than in the humid ones. For one thing, what is dry country now might have been rainier in the past, and in some cases is known to have been. Also, aridity aids preservation. In warm, humid regions the deep, unstable soil cover, rain washing and rapid erosion, thick vegetation camouflaging whatever lies beneath, all decrease the likelihood of finding well-preserved clues to the past.

Tools fashioned from resistant rock are the most enduring clues of early human development; they can remain little changed over millions of years. Bones that have become mineralized or petrified (converted to rock by groundwater action) are also very durable.

We can merely speculate about what was happening to Man's precursors as the upper and middle latitudes felt the chill and as glaciers were growing. Their habitat was switching from trees to the ground gradually while slow changes were taking place in their environment and bodies. One reason for their descent from the upper stories may have been simple curiosity. Being perceptive, they surely noticed there was action on the first floor as well as in the attic. But a more compelling reason, according to popular theory, was the changing climate associated with the developing Ice Age. The new highland barriers and changing atmospheric and oceanic circulations were upsetting old climatic regimes. The tropical rainforest zones diminished. Even where the tropics and subtropics did not feel a temperature drop, precipitation amounts and distribution were modified, markedly in some areas. Many always-wet areas began having a dry season. In forests that suffered a seasonal moisture deficiency the canopy of touching tree crowns thinned as some trees died. The forest's outer fringe changed to savanna, an open parkland of widely spaced trees and brush scattered over a grassy floor. Such was the case in Africa. Also, much of that continent (and parts of other continents) changed to desert or semi-desert.

The primates who found their forest habitat breaking up had to drop to the ground often to travel from one clump of trees to another, and they may have had to make farther trips than before for water.

There were good reasons for them to walk erect, as certain of them did, at least part of the time. Big cats and other predators lurked in the brush and the high grass, which grew as tall or taller than the characters in our story, most of whom were quite short. It was necessary for them to stand erect and peer over the top of the vegetation when venturing into such dangerous countryside. But into it our characters had to go. When foods of the old forest were no longer sufficient they had to use new foods. While they had been mainly vegetarian in the past they became increasingly omnivorous out of necessity, able and willing to eat just about anything. Like Australian aborigines of recent time, they probably consumed worms, rodents, and lizards. They may have eaten some meat as scavengers, taking what had been killed and left by animals; and they may have caught small and young animals themselves. The need to carry food back to the safety of a tree may have been another motivation for walking on only two limbs. As time went on they would become accustomed to non-plant foods of the parklands and grasslands and spend more and more time on the ground. With meat in their diet they then could go longer between "meals" and could live in regions drier and colder than those required by plant-food gatherers.

However it happened, changing environment of the oncoming Ice Age freed Man's progenitors from their old ways. The tree refuge was not always at hand, and methods of defense and attack had to be found. Cunning, alertness, and improvisation were requisites for survival in the new wide-open, killer-infested environment. Our displaced heroes and heroines were tempting meals for the larger carnivores and they needed a means of protection. When challenged by an attacker what better and more instinctive reaction than to grab a fallen branch (a familiar thing to grasp) or a large, long bone (a similar shape) and brandish or hit with it? Or what else might be lying on the ground? A rock. How neatly it fit the hand, and how well constructed was the long, loose arm for hurling it. Erectness was becoming, or had become, a normal position.

These dawn-of-the-Ice-Age characters may have been defenseless but they were not helpless. They had sharp vision, excellent eye-brain-arm coordination, and an already significant intelligence advantage over animal adversaries. They would learn by experimenting how to have the equivalent of a hoof or horn or claw or fang tooth by holding different objects in their adjustable, vise-like hand, and how to strike from a distance by throwing. They still had ape characteristics but they had a stance, a perception, that said they were more than apes.

Sometime during the Tertiary the branches of primates leading to modern apes and to Modern Man had diverged from a common ancestral line, and before the start of the Ice Age they were well separated.

As the Ice Age pageant unfolded, members of the ape branch would remain only bit players. But the others, they would steal the show!

CHAPTER 3

THE BEGINNINGS OF GLACIERS AND MAN

How often, if we learn to look, is a spider's wheel a universe, or a swarm of summer midges a galaxy, or a canyon a backward glance into time. Beneath our feet is the scratched pebble that denotes an ice age, or above us the summer cloud that changes form in one afternoon as an animal might do in ten million windy years.

LOREN EISELEY

THE ICE AGE is a many-mysteried time, and one of its most confounding mysteries is simply its timetable of events. Only near the very end of that epoch do we have dates that seem fairly reliable. For the start, middle, and, in fact, most of the Pleistocene there is not one positive date to go by. Many well-informed Ice Age specialists use few if any dates when discussing all but the latest Ice Age events. Contrariwise, some interpreters of the past express themselves all too confidently about Ice Age chronology although dates they cite may be disputed or rejected by others.

When you hear contradictory pronouncements about the Ice Age do not think that the confusion is yours alone. In any profession controversies smolder over theories, definitions, and interpretations. Among Pleistocene scholars the smoke really billows and the view is cloudy indeed. This is no secret. The British glacial geologist J.K. Charlesworth has written: "Perhaps no geological period has so divergent views as has the Pleistocene. Indeed, Quarternary geologists have long enjoyed the unenviable reputation of being among the most disputatious."

Keep an open mind and appraise new pieces of evidence as they come in. They are being assembled rapidly from all parts of the world. Take a wait-and-see attitude about Ice Age concepts whether they are old familiar ones or shocking new ones in the headlines. The coming years will bring exciting revelations about Pleistocene climate and geology and Early Man. With so many eager, well-trained men and women searching in places never examined before with powerful and intricate devices, breakthroughs must be near; and even the littlest finds help fill in the spotty jigsaw-puzzle picture.

If we are vague in our Ice Age timetable that is because to be otherwise would be misleading. But even though we are groping in the past we shall not be entirely in the dark. We have a tape measure with which to gauge past time, and although it may have to be shrunk or stretched later to adjust to new findings, it will help us space events in the uncalibrated millennia behind us.

The first date we should like to establish is the boundary between the Tertiary and the Pleistocene. We should like a clear-cut separation so we can say, "Here the Ice Age began." But it will not be so. One reason is that succinctly expressed by geographer Thomas M. Griffiths: "Just as nature is said to abhor a vacuum, nature abhors a boundary."

Scholars from many fields pool their findings to try to define the start of the Pleistocene. Botanists use plant fossils and preserved pollen as indicators of vegetational and climatic change. Zoologists study past changes in assemblages of animals as determined from fossil remains. (The first appearance of certain newly evolved animals helps define the start of the Pleistocene: the modern horse—this genus including the donkey and zebra; the modern elephant; true cattle; and, in parts of Asia, the true camel.) Oceanographers study cores of sediments pulled from the ocean floor. Geologists and geographers read the stratigraphy of glacial and pre-glacial deposits—as well as just about everything else on land and in the sea.

Any one of the many proposed Tertiary-Pleistocene boundaries actually covers a wide time span. As the Ice Age began, the changes that occurred, of whatever kind, did not arrive everywhere simultaneously and with equal intensity. The cold did not "hit" with the abruptness of a cold front passing, but took longer just to arrive than the entire duration of the Ice Age. Vegetation changes can serve as chronological reference points but plants take considerable time to readapt and form new communities. Using animal assemblages as time dividers also is indefinite because any given species did not arrive—abracadabra—everywhere at once, but evolved first in a certain area and then over hundreds and thousands of years multiplied and moved into new areas. That includes the human animal. If a fossil bone of a certain type of Early Man is found in one location and is tentatively assigned a date, we cannot conclude that that kind of Man "appeared" on Earth at that time. All we know is that he was there then. His kind evolved earlier, perhaps some place far away.

Considering the many complications, how do we arrive at any prehistoric dates at all, even approximations?

One way is by using a relative time scale. Earth scientists estimate the passage of time by determining from reliable historical records and their own observation how long it takes natural processes to make changes on the earth, and then they figure how long it might have taken such processes to have had similar effects in the past. They see how long it takes a forest to re-establish itself, a mature soil to form, a pond to fill with sediment and decayed vegetation, a stream to wear back a waterfall or carve a valley or build a delta; and how long soil leaching or rock weathering or groundwater action must go on to produce certain effects. Such methods give only a relative estimate of time, however. Numerical dates are sought too.

There are many methods of "absolute" dating in use, but they are not as precise or reliable as researchers would like. Some that are especially helpful in Ice Age research are the radiocarbon, potassium-argon, and fluorine measurements.

The unstable radioactive carbon-14 isotope is produced in the upper atmosphere by cosmic radiation. All living things assimilate carbon and contain a small amount of this isotope, and when a plant or animal dies the carbon-14 in it disintegrates steadily at a rate of one half the total in about 5,730 years. In other words, in 5,730 years about half the carbon-14 will have disintegrated, in another 5,730 years half the remainder will have disintegrated too, and so on. This test can be applied to any plant or animal substance, like wood, soil humus, bone, shell, or antler. Radiocarbon dating has thrown much light on the last part of the Ice Age, but it can supply dates back to only about 40,000 or 50,000 years ago with fair reliability. The dates must be used with caution. One reason is that if cosmic radiation, which produces the critical isotope, varied in intensity in the past, then the arithmetical answer is not valid. And it did vary. The half-life figure was revised in 1961, meaning that all prior radiocarbon dates require a small correction, and it could presumably be revised again if new information should call for further adjustment. Therefore, time spans based on radiocarbon dating should be read as "radiocarbon" years, not regular "calendar" years.

The potassium-argon method reaches farther back in time to many millions of years ago and promises to be a valuable aid in dating the start of the Ice Age and Early Man's origin after the technique becomes more trustworthy. The radioactive isotope potassium-40, which is a common substance in igneous rocks used for this test, decays at a known rate from the time the rock forms, and among its decay products is argon-40, a gas. If rock material can be found in which the argon was completely trapped as the breakdown occurred, then the ratio of potassium-40 to argon-40 gives an estimate of the age of the rock and thereby of fossils associated with the rock. However, if any of the argon gas escaped, the determined age of the rock will be too young.

Neither of these radiometric dating techniques, nor any other, can supply dates that are acceptable at face value. Every date needs verification by other means, and then it still carries a question mark. Radiometric dates are always stated with a plus-or-minus number of years to allow for statistical error. The test processes are complicated and tedious, and human error is an ever-present possibility. Error can result from other causes as well. In radiocarbon dating, for instance: An unearthed fragment of bone or wood did not lie in a sealed box all through the years. Soil, organisms, roots, and other matter that could affect the dating have been in contact with it; groundwater has dissolved and carried away material that should figure in the measuring and brought to that spot material from elsewhere. Just a speck of contaminant throws the computed date off. Contamination can occur even while the sample is being transported to the lab, or in the lab.

It can be foreseen, though, that as absolute-dating techniques are perfected they will reveal enlightening and startling results.

The fluorine test, which measures the amount of fluorine in substances that have been buried in the ground, is useful in determining the relative age of bones. Knowing the amount of fluorine in the groundwater of a given location, and the rate at which fluorine is absorbed by bone, one can judge how long a bone has been buried there. It is especially helpful in ascertaining whether all bones in a certain deposit are of the same age. This test will not tell time with precision, but it can be used to compare the fluorine content of a bone of known recent burial with that of a bone you want to date, and thus give a general idea of relative ages.

The personal element, however innocent, does tinge research. A researcher at times cannot help leaning in his interpretation of data toward the answer he wants to find. And there is another tendency of which we have to be aware because it bears upon Ice Age dating. It is the desire to find earlier and still earlier beginnings of things, as though the oldest thing found is a record, and it is an achievement to break a record. Earth's estimated age has been constantly lengthening, the start of the Ice Age is pushed farther and farther back, and the entrance of Early Man is likewise moved to an ever earlier date. This came about as research methods greatly improved. Although some of the oldest dates are not universally accepted, those extreme figures do recalibrate the tape measure if only because they have to be taken into account in generalizations and averages. Will the trend someday change, shrinking the tape measure, requiring us to shorten our time scale? Or are we in for still more staggering surprises about the antiquity of our world and the age of its people?

A few years ago William T. Pecora and Meyer Rubin of the U. S. Geological Survey in a coauthored paper, "Absolute Dating and the History of Man," presented this frank commentary on methods of dating the past, along with an apt warning in the last quoted sentence:

Our crying need for absolute dating of geologic events or materials has spurred such massive efforts by geophysicists during the past two decades that the activity along the path has raised, on the one hand, a host of geologic believers whose hope and faith are so frenetic that they have become uncritical, and, on the other, a host of geologic doubters who refuse to accept the probability that many gems lie within the handful of "number" grains offered them.

. . . All applicable dating methods and all possible materials are being used. This shotgun approach is not thoroughly satisfactory but is recognized as part of the evolution of the science. The possibility that all methods used today are wrong must be acknowledged.

So here we are again. When did the Ice Age begin? Shall we say it began when the Tertiary world felt the first alien chill? Or when the heavy snows first began to fall? Or when the tiniest glacier first became visible around the crest of some mountain? Or when mature glaciers started to move in unison? There we may have already covered the gamut of tens of millions of years. Even if we did have an accurately calibrated Ice Age time scale, we still would not know where on it to mark the start of the Ice Age.

Some glaciers formed in cold, snowy mountains while climate as a whole was still mild, before the Ice Age proper got under way. In Antarctica glaciers existed many millions of years before the time the Ice Age was first thought to have begun, maybe even in Early Tertiary; but just how far back in the Tertiary they did begin to appear, or how extensive they were in those early years, we do not know. So instead of saying the Ice Age began when glaciers appeared, we hedge and say the first glaciers formed sometime in the Tertiary and we hold off the official opening of the Ice Age until an indefinite time when glaciers became larger and more widespread throughout the world. This partially explains why one reads different figures for the duration of the Ice Age.

The old classical glaciological literature has glaciers going through four major expansions during the Ice Age. Each expansion is known as a glacial *stage*. Each major warm period between stages is called an *interglacial* interval. Of the three interglacials the second was believed to have been the longest, with the first and third considerably shorter. If the comings and goings of the ice had really been that methodical we would have a handy program for our Ice Age drama with four acts and three intermissions, but it was later learned that the ice's oscillations were not that regular and clear-cut.

In the Alps, where a four-stage pattern was first observed, it was originally outlined like this:

WÜRM STAGE (youngest)
Riss/Würm Interglacial
RISS STAGE
Mindel/Riss Interglacial
MINDEL STAGE
Günz/Mindel Interglacial
GÜNZ STAGE (oldest)

That nomenclature was derived from four streams in the Alps in whose valleys glacial deposits of different ages were distinguishable. Later in northeastern North America geologists likewise—by suggestion?—found four layers of glacial deposits, and these they have assigned to four stages according to age:

WISCONSIN STAGE (youngest)
Sangamon Interglacial
ILLINOIAN STAGE
Yarmouth Interglacial
KANSAN STAGE
Aftonian Interglacial
NEBRASKAN STAGE (oldest)

(As a memory aid, note that the most recent stage in both outlines starts with "W" and that the Alpine stages have an alphabetical order, from oldest to youngest.)

After four stages were originally identified in the Alps, geologists tried to make glacial deposits throughout the world fit a four-stage pattern, but that did not work. Still, there is a worldwide pattern of several cold periods with pronounced ice advances separated by warm periods when climate reverted to Tertiary-like conditions.

It would be convenient if the number and timing of all stages and interglacials around the world would match up synchronously but they do not. Due to the separation of the glaciated areas no way has yet been found to tell if or how glacial and interglacial periods in one place match with those elsewhere. To complicate the picture, within each major stage there were fluctuations of warmth and cold. If we plotted regional climatic variations of the Ice Age on a graph we would not have a smoothly oscillating line with well-defined cold dips and interglacial humps, but rather a complicated, irregular wavy line on which at places it is hard to tell whether a fluctuation is major or minor and whether a trend is on the way up or down.

Each time the ice expanded it must have done so by making a series of preliminary minor thrusts separated by setbacks as it advanced toward its climax. In other words, its spread was not steady, but a fluctuation of advances and retreats, like waves rolling onto a beach as the tide comes in. A wave rolls up and then falls back followed by another incoming wave which reaches up a little farther. Each successively higher wave washes away the outline left by the preceding, shorter wave, so how would we know the shorter wave had been there unless we had seen it? Only by knowing the habit of waves. We know little about the habits of the continental glaciers of the past. Did the ice rise to its high tide in the same oscillating way? We cannot say, but we assume so.

Each advance of the ice erased or disturbed much or all of the work of each former advance, but not always as completely as one wave erases another's mark. Sometimes an ice incursion destroyed the previous markings beyond recognition, but at other times evidence of earlier advances was left. Let us look at the case of the large Laurentide ice sheet (named for the Laurentide Hills of Canada along the St. Lawrence River) which covered northeastern North America.

It formed and spread out of Canada several times and at its southernmost positions reached approximately the courses of the Ohio and Missouri rivers in central United States. But in each stage the ice reached somewhat different limits, so layers of glacial deposits (gravel, sand, clay, and boulders) of different ages fringe the glaciated area along its outer margin.

The layer left by the last ice sheet, that of the Wisconsin Stage, is, of course, completely visible because it is on top. But the Wisconsin ice sheet did not everywhere cover the deposits left by earlier ice sheets. Protruding out from under it, but only in places, is a layer of older Illinoian deposits. Protruding in other places is a still older layer, the Kansan. The oldest layer is the hardest to find and identify. It is most deeply buried and was subjected longest to weathering and change. But it is encountered here and there in road cuts, excavations, borings, and in limited areas where upper layers have been eroded away.

It is not possible to map the complete outline of each layer of glacial deposits because every new incursion of ice reworked and covered the old surface it passed over. The sequential stratification of glacial deposits reminds one of a skirt with a number of uneven petticoats underneath. The Wisconsin deposits would be the completely visible skirt on top. The layers deposited during earlier stages or substages are various ragged petticoats that droop out here and there around the skirt's hem or show in holes worn through the upper layers. We cannot peek underneath to count how many petticoats there really are, or to see exactly where in the sequence a certain sagging petticoat belongs. There are some petticoats that are everywhere shorter than the skirt, as a good petticoat should be, and do not show at all.

So how can one say definitely that there are four clear-cut stages?

And how can one perceive when the Ice Age began? The beginning is buried.

Just as we cannot say when the ice began its Age, we cannot tell when advanced primates stepped over a threshold to become Early Man. Anthropologists suggest various criteria. Some would call a primate a Man when he stands completely erect. Some, when he starts to make tools. Some, when he uses fire or starts to speak.

When our tree-deserting ancestors reached the stage of using sticks and rocks that happened to be handy, they were still not Men. Present-day apes and some other animals, even birds, match such aptitude. Merely using picked-up objects as poking, smashing tools or as missiles was not a distinction from animals. If a primate *made* his tools, and especially if he did so according to a repeated pattern, then he is accepted as Man.

We can require that he have had speech to be human, but how do we know when speech began? When is a sound a word? And how many such sounds must he have made to have "speech"? Whenever it developed, speech helped Early People transmit ideas and cooperate. Let us not degrade them because they spoke with only a few sounds, if that is what they did. With those sounds and gestures they communicated as adequately, no doubt, in their simple world as we do in our complex one.

The beginnings of Man are eclipsed in the same penumbra that grays the beginnings of the Ice Age. It is clear that ice sheets and Early People were firming their footholds on this planet contemporaneously a few million years ago, although we are unable to draw time charts with straight lines across to show parallel worldwide development.

We have seen that the pre-glacial world was climatically genial, luxuriantly garbed in greenery, generously stocked with a grand assortment of animal life. Had it been less Edenic the change to Ice Age conditions would have seemed less severe. The descent from the warm Tertiary to the icy abyss was to bring more striking geographic changes than a drop from present conditions to Ice Age level would, for we are part way between the extreme conditions now.

As the Ice Age began, you recall, land was rising in many parts of the world, and new highlands had a pronounced effect upon climate. They invited snow. And the snow came. Delicate, fluffy, silent, gentle. Who could believe it was the infiltrating agent of a destructive force capable of changing the face of the earth?

Snow would fall wherever, whenever temperatures were too low for rain, mainly in colder regions toward the poles and on mountains. In such favored locales snow had been gradually increasing over innumerous years. Year by year, century by century, it was coming earlier in the fall and later in the spring, and the snowcover lasted for a longer time. In the early stages the snow would melt entirely away each summer; but later, traces would linger into fall. Once in a while in places where the snowfall had been especially heavy or the temperature unusually low, the last piles of snow, hidden in shadowy nooks, defied summer's warmth and were still there to welcome the first snows of fall. When a surplus accrued over a period of cold, snowy years, some snowbanks were able to persist even through an occasional warmer-than-average summer. During some warm periods all the snow might melt away and the foothold would be lost, but persistently and repeatedly snow would steal upon the scene again and do as before. Finally a lasting foothold was obtained. The next step was to secure that position and then advance.

Highlands in polar and subpolar latitudes were the most favorable places for glacier genesis. There the cold of high latitude worked together with the added cold of high altitude. Highlands provided the lifting mechanism required to cool air masses markedly and thus release heavy precipitation. Their height also guaranteed the coldness needed to make that precipitation be snow instead of rain and to keep that snow from melting once it had fallen. In a static situation where other variables of weather are not considered, temperature decreases with increased altitude at a rate of about 3.3° F. per 1,000 feet. A 10,000-foot mountain range would have a summit temperature about 33° F. lower than the temperature at its base—a sharp contrast.

Lower and warmer sites also could undergo snow build-up if snows were heavy enough. The deciding factor in whether snows would accumulate over the years was simply whether some snow could survive over summer and keep doing so.

The largest ice sheets, area-wise, would ultimately grow over Antarctica, northeastern North America, and northern Europe, in that order of size. And other sizable ones would form over northern Siberia, Greenland, and northwestern North America. (The *total* ice-covered area of North America, even excluding Greenland, was greater than the ice-covered area of Antarctica or Eurasia.)

In those high latitudes the summer sun circles or nearly circles the sky, and its blazing eye, though low, peers into valleys and niches even on poleward slopes. But the snow packs passed the summer crises, defied the sun's probing rays, and clung on. Fragile snow appears a weak rival to the omnipotent sun. But let us zoom in for a close-up and observe how soft, porous snow is transformed into strong, resistant ice.

Snowflakes take the shape of stars, hexagons, needles, columns, or other fine designs. These lacy ice crystals with feathery edges fall singly or in masses or fragments. Even as they land, warmth which would destroy them begins work. Many flakes melt immediately. But even when temperatures are low enough to preserve them the thirsty air is able to diminish them by evaporation. Their thin, fingery points are vulnerable. Evaporation increases as temperature rises, but air need not be warm to do a good job of evaporation. Cold air is commonly desert-dry and can readily absorb moisture. It is not even necessary that snow melt and change to water to be evaporated. By the process of sublimation solid ice is converted directly to gas, water vapor, without passing through the liquid stage. (You can observe the results of sublimation in your kitchen if you place loose, freshly broken pieces of ice in your refrigerator tray. They have sharp edges, but leave them in the freezer a few days and then notice how those edges are becoming rounded and smooth. The ice pieces have not melted but sublimation has been at work sucking away their exposed corners.)

This evaporation along with warming and softening of the fallen snow, or partial melting and refreezing, changes snowflakes from filigreed crystals to tiny spheres of ice. These little icy grains, known as *firn*, settle together more compactly than fluffy snow does. Then fresh snows fall upon them, cover and protect them, add weight, and press the grains still closer together. A new upper layer of firn forms from the fresh snows. The increased weight of more snow and more firn accumulating on top adds ever-increasing pressure to the older grainy ice below.

The firn grains attach themselves to one another just as loose ice cubes in the refrigerator tray grow together even though melting has not occurred. Pressure in the thickening firn fields facilitates this regelation, or freezing together, and the ice grains recrystallize, assuming new shapes. The grains join and form larger and larger crystals. Fresh firn grains have diameters of only about an eighth of an inch, while full-grown ice crystals in a mature glacier may be several inches across and irregular in shape.

So, in our story, over the years fluffy snow was converted to hard ice. The ice strengthened. Weight of new layers above compressed the lower ice more and more. Gradually most of the air among the crystals was squeezed out, but some was locked in pore spaces, there to be confined until the ice cracked or melted. That may have happened soon thereafter or not until thousands of years later.

These bodies of ice, strong as they were, were still not glaciers. A glacier, by definition, is a mass of ice that has formed naturally from compacted snow and that moves or has moved. (The firn and new snow atop the solid ice are also considered part of the glacier.) The ice bodies we have described were still stationary. But they were slowly taking form, about to be born from their stony wombs as living glaciers.

Those are the cold facts of life concerning the birth of glaciers. An embryonic glacier may be conceived, as has been described, wherever snows lie undisturbed for some time—in the embracing arms of a mountain range, a shadowy bed of hills, a protected plateau. The body of ice which the snow begets is, in a sense, the fetus of a glacier. It increases in size, but not until that growing body moves under pressure of its own weight does a true glacier come into being. Then it eases itself along the terrain's structural grooves, extends its small, weak limbs outward and begins to let its effect be felt upon its surroundings.

Now we have glaciers and the action is under way. Snowlines on mountain ranges drooped lower with the passing years. Communities of vegetation gradually re-established themselves where it was warmer, farther down the slopes of mountains or nearer the equator, relinquishing their old neighborhoods to hardier groups of plants which could tolerate the lengthening colder winters and changing moisture conditions. Animals too adjusted their range to stay with climate that suited them and food they were used to, whether they were herbivorous or carnivorous.

While glaciers were being born, Early People were being born, but their respective birthplaces and first areas of development were quite separate. Glaciers in their infant stages were cradled in cold climates, and with time they would creep outward from those regions toward maturity. Early People, on the other hand, supposedly originated in warm regions; it is believed in their natural state they would have found cold regions uncomfortable, if not uninhabitable. For now, these two remarkable creations of the Ice Age, glaciers and Early People, in their separate cold and warm arenas, are just fighting to survive. As our story progresses they will extend their areas of influence and eventually, inevitably, they will meet.

CHAPTER 4

HOW TO TRACK A GLACIER

The study of glaciers has its own special kind of interest and excitement; one is not only pitting one's mind against secretive Nature, but her sheer physical difficulties can form half the struggle.

J. D. IVES

WE CANNOT say which frosty peaks or cold plateaus were the gathering grounds where the first ice regiments rallied and began their drives. That may never be known, for glacial erosion which followed was so rough and prolonged that the initial battle scars were worn away. But there are clues that point to the places of origin.

It is easy to retrace the paths of short, localized glaciers that never roamed far from their isolated mountain bastions, such as those of the Cascades, Sierra Nevada, or the Alps. But the starting points of the granddaddy ice sheets, those largest of all glaciers which stretched up to a thousand miles or more from home base, are harder to locate. They began in northern North America, northern Eurasia, and Antarctica—somewhere. Where? Antarctica is still under wraps, 98 per cent ice-covered and not revealing much; but we are not as concerned about where glaciers first took root on it as we are about where they began in North America and Eurasia. Upon those two continents diehard veterans of the once-omnipotent ice army are still entrenched, and there energetic, technically advanced populations are multiplying and moving into unoccupied territory at an unprecedented rate. Upon those continents Ice and Man, both mobile and both with a tremendous potential of expansion and destruction, face each other across narrowing neutral zones. In many places they stand toe to toe. Something more than idle curiosity makes us ask: Where did the main ice armies have their beginnings and which way did they move?

How can earth-science detectives hope to answer that question? Fortunately for their search, the ice did not tread with dainty footsteps. Its conspicuous spoor shows what routes it traveled.

When glacial ice overspreads an area it leaves a carpet of drift. *Drift* is the general, all-inclusive term for material of any kind carried and deposited by glaciers and their meltwater. *Till* is another name given to glacier-deposited material, but it is material that was let down directly from the ice and does not include material deposited by meltwater.

The drift which the ice left is in some places thin or patchy. In other places it is hundreds of feet thick, coating plains, filling up valleys, and forming or reshaping hills. Drift can be a heterogeneous conglomeration of stony material, comprised of boulders of all sizes along with gravels, sands, and silts mixed in any proportion or combination. Or it can be uniformly graded, neatly sorted by meltwater, so that materials of the same fineness or coarseness are deposited together as though passed through sieves. Meltwater that drains from ice meticulously works over its load just as running water anywhere does. You can watch moving water sort sediment on your sidewalk or in gutters after a rain, or in your garden as you water your plants. Meltwater sorts the glacial materials, dropping those of like weight together. Heaviest rocks drop out first near the ice, while the finest particles can be carried farthest and with least velocity. Glacial material laid down by meltwater is known as *outwash*. It may be spread by diverging or crisscrossing braided streams to build delta-like fans or wide *outwash plains* in front of the melting ice, or it may be channeled along a river valley as a *valley train*. If piled as a mound alongside the ice by a stream running off the glacier, it is a *kame*.

Since the ice sheets formed and expanded several times, there are, as we said before, several layers of drift, one atop the other. Drift layers of different ages can be distinguished from one another in places by their composition, color, and degree of weathering; by layers of old soil that formed on top of the drift during intervening warm periods; or by datable material within the drift. We cannot know where the first ice sheets on each continent started because their traces have been buried or scoured away by later ice sheets.

The best we can do is assume that the earliest ones followed a growth pattern similar to the last ones of which we have the clearest picture. It seems that the recurring ice sheets grew out of approximately the same source regions on their respective continents each time and moved in somewhat the same directions, for their drift layers—the ones we can identify—overlie each other and have rather similar shapes. That is not to say that the ice came consistently from the very same locations during all stages or throughout any one stage. In North America, for instance, sometimes a surge of ice would come from west of what is now Hudson Bay and at another time from east of the bay.

Each sheet seems to have spread like a pie crust being rolled out, splitting into lobes along the edges, growing more easily in some directions than others. Pie crusts or ice sheets—each one that spreads out has a shape all its own. But although the sizes and shapes of successive ice sheets differed, they seem to have grown along similar lines and in their farthest thrusts reached, or were reaching toward, the same approximate limits.

On a glaciated surface the direction of former ice movement can be determined by several methods. One is reading *striations,* or scratches, cut in bedrock the ice passed over. Where they occur they run generally parallel like sweep marks from a giant broom, indicating direction of movement. Fine grit in the moving ice etched thin, shallow lines which soon weathered away. But large, sharp rocks gripped vise-tight in a glacier's base gouged bigger gashes, in some places several feet deep and up to hundreds of feet long, straight and uniform as though a mighty precision machine had grooved them. Striations are more pronounced on hillsides and outcrops that faced the oncoming ice, and are less pronounced on the lee, or downslope, sides. Striations have been carefully mapped and as one reads the overall pattern the composite picture of movement is seen and centers from which the striations diverge become apparent.

Erratic boulders supplement striations in blazing a glacier's trail. These are the largest, most obvious objects deposited by glaciers and one of the best indicators of their direction of travel. Their location leaves no doubt which way the ice went, for glaciers cannot move backward. (When we say the ice withdrew, or retreated, of course, we do not mean that the ice itself went backward but that the outer edge was melting back faster than the body of ice was moving forward.)

Over much of the glaciated surface erratic boulders are strewn as generously as poppyseed on coffeecake. Before their origin was known they were called "lost rocks" or "foundlings" for they were strangely out of place. Europeans of previous centuries used supernatural and fictional explanations to account for erratics. It was said they were playthings of mythical gods or giants, or were cast there by the devil or witches, or fell from a comet's tail. No one had yet imagined an even more incredible explanation—a traveling superstructure of ice. Many erratics are so huge that nothing but the ice, nothing else natural or man-made, could have transported them. One of Europe's most famous erratics, Pierre à Bôt (Toadstone), is a granitic block that was carried about 70 miles from Mont Blanc north to the Jura Mountains. It weighs an estimated 3,000 tons and is as large as a house. In America, New England has some whoppers of 5,000 to 10,000 tons. South of Calgary, Alberta, there is a quartzite erratic, now broken apart, which measured over 150 feet long and was estimated to weigh 18,150 tons.

A geologist who knows his area well may examine a rock with distinctive characteristics and tell where it came from, the way the FBI can pin a name to a fingerprint. For example, metallic copper from Upper Michigan, or a block of a certain granite from Westmorland, England, with its special kind of pink feldspar crystals, is as easily retraced to its home as to a mailing address. A boulder of Precambrian crystalline rock from Ontario set on a limestone ridge in Ohio is as conspicuously out of place as a polar bear on a Cleveland sidewalk. If a rock can be identified as having come from a certain outcrop to the northeast, for instance, then it is known that the ice moved in a southwesterly direction over that outcrop. A rock need not be huge nor have traveled far to show what direction the ice carried it. The farthest that erratics are known to have been carried is about 800 miles.

The ice picked up chunks of rock, dragged them along, and dropped them helter-skelter in its path. Soft rocks crumbled easily under the ice's grinding and crushing action. But more durable ones like gneiss, granite, quartzite, basalt, gabbro, rhyolite, and marble withstood glacial mutilation better. If carried deep in clean ice and not subjected to abrasion, rocks could be dropped without having

been worn much. But most were subjected to rough handling which
scratched them, wore off their corners, smoothed their sides, and
left them partly rounded. Some that were frozen rigid in the ice's
base, like a gemstone fixed in a faceting machine, were worn flat on
the side that took the grinding as ice with great pressure scraped over
its raspy bed. The ancient crystalline shields of eastern Canada and
Scandinavia are composed of some of the oldest, hardest rock on
the planet, and they supplied outward-moving ice with an abundance
of tough, resistant boulders that could travel long distances without
losing their identity. Thousands of years later when geologists, ex-
cited with the first imaginings of vanished floods of ice, took to the
countryside seeking tangible proof to substantiate that wild idea,
those durable erratics were standing as indisputable evidence where
they had been dropped.

Not all erratics can be tracked back to their source. If sources could
be easily located more geologists and rockhounds would be million-
aires, for scattered among the mix of glacial drift are gold and diamonds and other precious stones and minerals. Southern Indiana's
glacier-built gravel hills were once so rich in gold nuggets that commercial recovery was considered, but most of the gold was extracted
by individuals panning. One prospector panned successfully for about
fifty years.

Diamonds have been found in the drift of Wisconsin, Indiana,
Ohio, and Michigan. The source of these valuable erratics has been
hunted for but never located. Somewhere to the north the buried
pipes of an old worn-off volcano, or of more than one volcano, probably still hold a hidden fortune of these gemstones. Since diamonds,
the hardest of all natural substances, can withstand tremendous
grinding and travel great distances without disintegrating, the volcanic source from which they came may be quite distant from the
points where they were deposited. Judging by the paths of the different ice lobes in whose drift the diamonds were found, the source
was somewhere between Lake Superior and Hudson Bay. Rough
diamonds as large as 16 and nearly 22 carats were discovered in Wisconsin's drift. Glacial gravel pits may contain diamonds, and some
undoubtedly are lying in the gravels of roads and parking lots, but the
chance of finding one is slim due to the overwhelming preponderance of other rocks.

The airplane affords a wide panoramic view of the distribution and relation of glacier-made features, providing another method of ascertaining the ice's direction of travel. Viewed from the air the ice's tracks over some large areas are as plain as a plow's furrows. In still-natural places like central Canada, and even in some cultivated farm country, one can look down on glaciated land and see a directional grain: elongated, scoured hills and valleys; long, narrow lakes; drift molded into oval, streamlined hills oriented in a consistent direction; everything smoothed along lines parallel to the ice's path. (It is a bit shocking to think, while you are flying over such glaciated country in a small airplane, studying terrain, that if the ice sheet still existed you would not be looking down on it as you imagine yourself doing, but you would be frozen *in* it, plane and all!)

Landforms built by the ice and its meltwater tell much about how the ice made its final retreat, and one might consider its retreat an illustration of its advance in reverse, at least in a general way. Composition and arrangement of the drift, including outwash, reveal the manner and direction of the ice's fluctuating withdrawal. They show, among other things, what shape the ice sheet had and at what positions the ice margin halted during retreat. Reviewing the ice's recession is a little like seeing a film of its advance run backward, though of course it could not be entirely so. Places which were most favorable for the glaciers in their formative years seem likely to have been most favorable in their waning years too. The glaciers in old age pulled back to protected, cold, snowy retreats, in many cases probably the same places where they were born.

Not all places that support glaciers today were source regions of the largest-size glaciers, the continental ice sheets. Some glaciers in spots of limited snow accumulation grew just a little larger than they are now and that is all. But others situated in areas of greater snow receipt swelled to enormous dimensions.

Not all our present glaciers are continuous holdovers from the Ice Age either. During the post-glacial warmup, temperatures were higher than now, and in many places where glaciers—particularly small ones—exist today all ice disappeared at the close of the Ice Age; the glaciers now there have re-formed in recent centuries.

Utilizing various types of evidence—the pattern of repeated ice expansions, striations, erratic boulders, arrangement of deposited features, manner of ice withdrawal—we can point to places on a map where the Pleistocene glaciers probably originated. We could infer that if ice sheets were to grow again they would start about where the last ones did.

The reconstruction of the Ice Age which follows assumes that geographical processes at the onset of the Ice Age and throughout it operated substantially as they do now. This may not be true. But we do not call upon any cataclysmic geologic or climatic disorder to explain what happened, for we do not know that there was any such occurrence.

Some change did take place to trigger the start of continental glaciation. In sundry theories that change has been attributed to both ordinary and extraordinary events. Later we shall weigh various theories of the cause, including some that tax the imagination. The cause of the Ice Age may, indeed, have involved the fantastic. Or it may not. All that is needed now to initiate another ice expansion is something as ordinary as a drop in the average temperature of just a few Fahrenheit degrees, or a modest snowfall increase or topography change in strategic locations, and time to let the effect take hold. Bear in mind, as we recount how an ice sheet came into being and how it spread, that if this is how it happened this is how it could happen again.

CHAPTER 5

THE BATTLE OF THE ICE AGE BEGINS

. . . Hot, cold, moist, and dry, four champions fierce,
Strive here for mast'ry.

JOHN MILTON, *Paradise Lost*

WE LEFT the young glaciers just as they had begun their march, tiny insurgent guerrillas in scattered outposts beginning to test their strength. Cold guarded them, snows provisioned them, and they grew in size and power. As long as snows kept them well equipped nothing on Earth could stop their impending drive. Their arch enemy, the only force that could ultimately halt them, was the distant sun.

The interest level and importance of any contest—be it in nature, sports, politics, business, war, or anything else—is measured by the stature of its contestants and the prize at stake. The prize here was a world. The contestants were two giants: the Sun, energy epitomized; and the Ice, mightiest force ever to cross the land.

The sun had been attacking the gathering snows and forming glaciers back in the Tertiary. It attacked not only by its direct rays but also indirectly as its warmth was in the winds, the water, the soil, the rocks.

This was a to-and-fro, pendulum-type confrontation. At times the sun's warmth had the upper hand and glaciers retreated. At other times glaciers made gains. There were countless fronts, all along every margin of every ice body. The ice was forced back along some fronts while making headway along others. Now as the Ice Age was getting under way most ice fronts were advancing.

As we look to places where the glacial armies mustered we picture highlands in cool, moist regions. Elevations and details of topography have changed since then. Prominences wore down considerably during long, intensive ice scouring. But not all elevations succumb easily to erosion. Some are composed of highly resistant rock. Also, gradational processes which would level the land are counteracted by tectonic processes which uplift it and keep it irregular. Subterranean pressure can be raising mountains, maintaining their high altitudes, even as erosion rips off their tops. Such is the case now in the Himalayas and other young, rising mountains.

In Europe the main snow-gathering region seems to have been the high mountain-and-plateau spine of the Scandinavian Peninsula. Small ice caps and glaciers exist there now. Westerly winds coming off the Atlantic, their moisture-holding capacity enhanced by crossing the warm ocean current, bring abundant snow to this region. Secondary areas of snow accumulation in Europe, where moist air masses had easy access, were most of the mountains of that subcontinent, especially the Alps and the highlands of the British Isles. Also Iceland.

Broad-shouldered North America, snowiest of the continents, was to bear the heaviest load of ice. Greenland probably held some of the earliest glaciers. Other original gathering places of ice may have been mountainous Labrador and the northern Quebec peninsula, the mountainous Arctic islands, and later the lower area around Hudson Bay. Hudson Bay was not a bay before glaciation. Then it and the land around it stood higher. The ponderous weight of ice that lay on northeastern North America for thousands of years depressed the rock basement and this area has not yet entirely rebounded to its former elevation.

Baffin Island, largest island of the Arctic archipelago, located just north of Labrador and west of Greenland, contains two impressive ice caps now and many additional glaciers and icefields. Greenland's other western neighbors, Devon and Ellesmere islands just north of Baffin, contain considerable ice too. Greenland might well have resembled those partially iced islands while its ice sheet was forming. First there were small glaciers, individual gems set like solitaires in its mountains; and ultimately they grew and coalesced into one large, rounded cabochon jewel mounted in the prongs of the encircling mountain rim.

Another area of early glacier build-up must have been North

America's western "armpit," the curve of coastal ranges that runs along or near the Pacific coast of Alaska and Canada and includes many of the continent's loftiest peaks. This west-coast region is a counterpart of Scandinavia in that it has a mountain barrier lying in the path of moisture-laden Westerlies which have passed over a warm ocean current.

So in North America the glaciers probably formed near the oceans first, not in the colder interior of the continent. Strange as it seems, the lowland interiors of Canada and Alaska, in spite of their low temperatures, were probably meagerly supplied with snow just as they are today and so were unlikely places of glacier genesis. For the same reason Siberia was slighted glacially. Although Siberia has recorded the lowest temperatures outside Antarctica, its receipt of snow is relatively small. In winter, where sun angle is low, nights long, and marine influence weak, the land surface becomes extremely cold and creates a strong high-pressure system with outblowing winds (the opposite of the inblowing sea-to-land circulation of the summer monsoon when low pressure dominates over the heated land). Asia's winter outpourings of cold air masses are stronger and more persistent than North America's because it is a larger land mass. Moist air masses from the warm ocean find it difficult to move to the continent's interior over mountains or against the prevailing Westerlies. Siberia was not an especially good place for glaciers. It has small ones now and has had larger ones. There was a good-sized ice sheet over part of Siberia but it was inferior to the Big Three of America, Antarctica, and Europe in thickness, area, and vigor. The extent of Siberian glaciers is not clearly mapped, for systematic glaciological field work began much later in the USSR than in Europe and America—not until the 1930's.

Antarctica, which today supports the greatest mass of ice—about 85 per cent of the world's total—must have been one of the first places to support glaciers. This was a well-watered region with plateaus and mountains rising up to catch moisture. It seems to be more than one island. Soundings made through the ice to bedrock indicate that parts of the so-called continent are below sea level. Some below-sea-level depressions are caused by ice weighting them down, but others undoubtedly were ocean passages between islands before the ice covered them.

How easy it was for glaciers to form there at the hub of the Southern Hemisphere whose surface is 81 per cent water. There, where a superabundance of moisture was received from the wide surrounding seas, where the sun is excluded for up to half the year and weak the rest of the time, the advantage in the conflict between ice and warmth was with the ice.

The polar area of the Northern Hemisphere is a reverse image of the Southern in that there an ocean is centered on the pole and it is surrounded by large land masses. But lack of land at the north pole was no handicap to glacier formation in that hemisphere. While the Southern Hemisphere is glacier champion now with Antarctica's unequaled volume of ice, the Northern was champion in the Pleistocene. Although it had no land at the pole, it had the largest land masses in the snowiest latitudes. Cold alone, as we have noted before, even polar cold, does not make glaciers. Moisture is the vital ingredient. And there is more atmospheric moisture away from the poles than at the poles, generally speaking, because warm air can hold more moisture than cold air can.

As we visualize the beginning of the Ice Age we should mentally remove from the uplands, where glaciers first stepped into the scene, the striking, ice-created features that characterize that terrain now, for in our story the ice has not yet done its work. We have to remove all the familiar, beautiful artistry of the ice's deft fingers—the jagged, pointy peaks; castellated and serrated ridges; scoured-out basins holding glistening tarns of clear water; sheer, sliced-off cliffs towering above the valleys; tumbling streams and cascading waterfalls plunging without channels over ice-steepened precipices, splashing from nowhere to nowhere. How the scene dulls to drabness as we erase these inspiring gifts of the glaciers.

Mountains and hills had been etched only by water with a touching up by the wind. Water does, by far, most of the sculpturing of the land, but it is a conservative artist compared to the bolder, stronger ice. If we compare the power of the three agents of erosion —wind, water, and ice—we might say that the wind is a featherduster, water a rake, and ice a bulldozer.

These were humid regions where the glaciers began, so the slopes of stream-cut V-shaped valleys must have held thick accumulations of rock and soil which had slumped down from the summits and sides and were waiting their good time for the streams in the bottoms of the V's to carry them away.

Build-up of glaciers began slowly. Most were probably in protected niches and hollows high among the mountains or up in the hills. They thickened and widened until pressure and gravity sent them out into the valleys.

As these young glaciers moved out from their starting points they picked up and carried with them pieces of rock and thus rounded out basins beneath themselves, as glaciers do today. These quarried basins, called *cirques* (pronounced "sirks") or *corries*, look like half bowls or like hollows made by giant ice-cream scoops. They are found around the tops of glaciated mountains. A glacier does much of its erosion by plucking. A cirque glacier plucks rock from its sides and bottoms by molding itself around loose pieces and pulling them along. Rain and meltwater run into rock joints, or cracks, and freeze, attaching chunks of rock to the main glacier body. When the glacier moves, these rocks are dragged with it. Ice has a powerful adhesive grip which you can feel when your hand is damp and you touch a freezing-cold object like an ice-cube tray. Rocks that break loose from a cirque's upper walls fall upon the glacier below and they too are transported away by the outward-flowing ice.

As our incipient glaciers kept picking up rocks and carrying them along they deepened and widened the cirques that held them, and thus were better protected in deeper, shadier pockets and had larger catchment basins in which to collect falling, wind-swept, and avalanching snows.

If a mountain peak had cirques biting into it from all sides the cirques as they enlarged would eat backward into the mountain. In time their sides would meet and be worn away and then only the back walls would remain, leaving a pointy central peak—a *horn*. The Matterhorn on the Swiss-Italian border is a classic example of a horn and a challenge to climbers because of its exceptionally steep slopes.

If cirques should be lined up in rows on opposite sides of a mountain range and erode backward into the divide until their backs nearly meet, there is left along that divide a high, slender, ragged wall known as a *comb ridge* or *arête*. Eventually that will be worn away too.

Cirque glaciers extended out like tongues over the lips of their basins and elongated themselves down the valleys which had been carved by the headwaters of streams. The lower valleys now carried the glaciers' meltwater. These were natural routes for young glaciers to follow. As gravity pulled, their bodies sagged into those outward-leading channels already there.

Ice moving down a valley is classified as a *valley glacier*. It flows smoothly except where there are abrupt drops or sharp turns or an uneven bed, and there its surface cracks open in *crevasses*. But as the flow evens again the crevasses heal. Some valley glaciers bulge over steep or step-like drops, the equivalent of waterfalls or rapids in a river, tumbling down in *ice falls*. On whatever ledge or slope they land they collect themselves and push on.

Small glaciers from the mountains inched out of numerous local pockets, slid slowly downslope in separate ice streams until they met other small glaciers coming from neighboring valleys, and then joined as tributaries of steams do, making wider, fuller rivers of ice.

While mountain glaciers are confined between valley walls they move the way their valley directs them. They have been described as rivers in slow motion, and in some ways they are. Like rivers, they move fastest where friction is least, so the surface moves faster than the bottom and the center moves faster than the sides because contact with the ground produces frictional drag. But a valley glacier, being a solid, is less limber than a fluid river and has more power and carrying capacity. Really it is not solid but plastic, for it molds itself to the shape of the land about it. Ice does not rely solely upon gravity for its movement as water does, but is pushed by a force within itself still not understood.

Like water, a glacier drains downward seeking the sea, but it has two special abilities that water lacks. It can erode below base level—that is, the level of the water body toward which it is draining. And it can flow upslope—that is, to the extent that pressure from behind pushes it up.

The ice sliding through the stream-cut, V-shaped valleys rose to heights the faster water had not touched even in flood, and plowed up vegetation and carried away talus and other rocky materials, using them as abrasive tools.

While a stream darts agilely around a bend, the slower glacier with sluggish directness and greater force grinds off the tapering spurs, making the valley sides steep, even perpendicular. The result is a new profile for the valley, a U-shaped trough. Where these valley-glacier troughs meet the sea they are *fiords*. The coasts of Norway, northwestern North America along the Pacific, southern Chile, and

New Zealand, all on exceptionally snowy west sides of continents in upper middle latitudes, have some of the world's most deeply fiorded coastlines. Fiords exist along other heavily glaciated mountain coasts too.

Glaciers are like great conveyor belts carrying stony cargo as they move, and valley glaciers are like "down escalators" transporting rubble downslope and dumping it at their lower ends. Some of the rubble is strewn underneath and along the sides of the glaciers throughout their full length, but the rest is delivered to the glacier terminus, where it is piled up as the ice melts. If the glacier ends on land and its front stays in one place a long time, a hilly ridge of mixed debris is built there. This *moraine* has the curved outline of the ice's tongue or lobe. If the ice advances farther it simply pushes the moraine material ahead or overrides it. A moraine left at the ice's farthest position is an *end moraine*. If the ice pauses in its retreat and builds another moraine behind the end moraine, that is a *recessional moraine*. Moraines usually consist of boulders and coarse materials because meltwater running through and away from a moraine leaves the heavy debris behind and carries the finer material farther down the valley. The escaping streams clog with gravel and sand and subdivide into many rivulets of less carrying capacity as they flow in braided fashion through the choking outwash.

Where the ice rubs against valley walls it forms *lateral moraines* composed of rocky debris scraped and avalanched from the sides. Where two tributary valley glaciers meet at a "Y" confluence the lateral moraines of each join along the merging sides and make a *medial moraine* which appears as a dark streak down the length of the glacier as the two ice streams flow alongside each other.

So much is known about valley glaciers because they can be observed in action. You can hike right along the sides of many of them and observe how they scrape the valley walls; you can even walk across the surface and see the rocks embedded in and lying on the ice. You can walk into smooth crystalline tunnels dissolved by subglacial streams and peer from the precarious brink of crevasses and holes into the translucent blue bodies. You can hear the ice crack under the strain of movement, hear stones rattling through glassy caverns below and the gurgle of little streams under and in the ice.

We know less about glaciers after they become larger and outgrow the valleys, but we can imagine how their expansion in the Ice Age must have taken place.

Eventually our valley glaciers reached a more open area where they widened into a rounded, pancake shape, becoming *piedmont* (foot-of-the-mountain) glaciers. Several spreading piedmont glaciers from side-by-side valleys would run together and form a "puddle" of ice along the mountains' base. Still they were dependent upon the higher elevations where they originated for nourishment that kept them growing and moving. It was as though there were a ladle above the mountaintops pouring batter over the landscape. The more "batter" poured, the farther the glaciers could flow and the thicker they could become.

Valley glaciers flowed down from the highlands like a gathering of the clans, uniting with each other into whole systems of interconnected streams. They thickened till they filled the valleys and intermontane basins to overflowing. Still higher they grew, drowning divides and mounting up over the very cirques and mountaintops that gave them birth and first nourished them. Thoroughly weaned now, the ice received snow over its whole wide surface which covered an entire mountainous region.

The consolidated glaciers had grown into *ice caps* and covered large areas completely, giving the aspect of a dome, dominant and secure. Now although the ice's movement was directed by buried valleys, the upper part of the dome was free to spread radially in all directions. Prominent topographic features could divert or retard the flow, but none could stop or control it.

On plateaus too, ice caps grew through the accumulation of snow of many years on the level surfaces.

In this still-early part of the Ice Age, ice caps crowned parts of Antarctica, the Scandinavian Peninsula, northeastern Canada and some Arctic islands, the Alaska-Canada Pacific coast, South America's southern Andes, and some highlands of Asia. Those ice caps destined for bigger things sent extensions into outlying territory like an octopus' tentacles reaching for prey.

Farther afield additional areas were being prepared for subjugation. There was a general lowering of the snowline, that critical altitude above which snow does not melt in summer and the year's average temperature remains below freezing. Above the snowline glaciers can form. They can descend below the snowline as long as the source region above supplies enough ice. Many highlands that had remained snowless now came under the falling tarpaulin of year-round white. First it covered just their summits, letting new glaciers encamp there, and then it drooped still lower covering the slopes.

When the snowline descended onto the wide lowlands the glaciers made giant strides. Ice on a level surface, not having the pull of gravity, must be about 200 feet thick to move under pressure of its own weight, and then it spreads outward. If, say, a foot of granular firn accumulated every year (the amount would depend upon how much snow fell and escaped melting) then in 200 years, if there were no setbacks, a moving glacier could be started, but accumulation must have begun much more slowly than that.

Nevertheless, as heavier snows built up layer upon layer over the broad plains, how swift must have been the ice's conquest. Ice floods from many highlands ran together joining the rising seas of ice on the lowlands, for centuries or millennia, until united they formed vast *ice sheets*, the largest type of glacier.

If the forces of warmth could not stop the marching glaciers when they were just weak lines of irregulars, what now could halt these growing hordes?

CHAPTER 6

THE CONQUERING COLD

O Wind,
If Winter comes, can Spring be far behind?
PERCY BYSSHE SHELLEY

GLACIERS are masters of both offense and defense. Even a small valley glacier gives a sample of ice's power of endurance and conquest. You are at the base of one in midsummer when it is weakest, when its front is splintered and soggy and shriveling back. Around the next hill the weather is balmy; but here, blowing down that ice-floored chute, is a wind so strong and bitter cold it stiffens your fingers and chills the marrow in your bones. You hunch up your shoulders and shiver and know that come winter that sickly ice front will recuperate and harden again into a solid phalanx, ready to charge ahead. If a weakened valley glacier can exhale a breath that frigid on a summer day, how much more did a gigantic ice sheet do in the longest, fiercest winter the world remembers?

Picture a young, healthy Pleistocene ice sheet in the process of expanding and watch how this mobile mass, temporary and vulnerable as it was, for thousands of years influenced the environment in its favor and fostered its own growth. This reconstruction of what happened is conjecture based on generally accepted facts about glacier growth and on the assumption that world geography then was basically the same as now.

The ice sheet was a featureless white tableland several thousand feet thick, fairly smooth and undulating on top, irregularly shaped and sloping down at its margin. Here and there mountaintops protruded like towers above a frosted citadel. It must have looked much

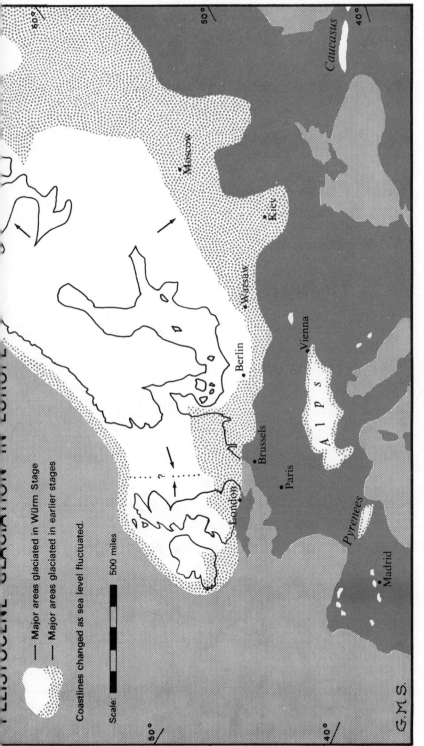

PLEISTOCENE GLACIATION IN EUROPE

— Major areas glaciated in Würm Stage

— Major areas glaciated in earlier stages

Coastlines changed as sea level fluctuated.

Scale: 500 miles

Caucasus

Moscow

Kiev

Warsaw

Berlin

Vienna

Brussels

Paris

A l p s

London

Pyrenees

Madrid

G.M.S.

PLATE 1.

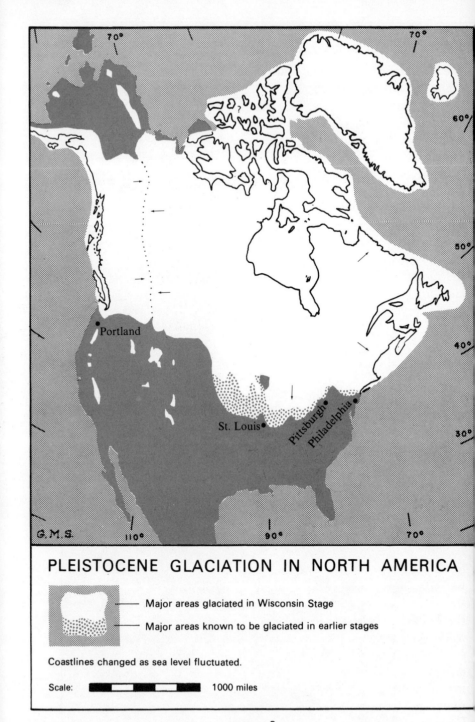

70° 70°

60°

50°

40°

30°

Portland

St. Louis•   Pittsburgh•
                Philadelphia•

G.M.S.   110°   90°   70°

PLEISTOCENE GLACIATION IN NORTH AMERICA

⬜ ——— Major areas glaciated in Wisconsin Stage

▨ ——— Major areas known to be glaciated in earlier stages

Coastlines changed as sea level fluctuated.

Scale: ▬▬▬▭▬▬▬   1000 miles

PLATE 2.

like Greenland's or Antarctica's ice sheet; but we cannot use them as examples, although they too tend to perpetuate themselves, because they are surrounded by ocean and cannot expand normally. We shall be thinking rather of those less confined ice sheets of Europe and North America which had wider spheres of operation and could spread out over the continents.

Continental ice sheets, since they have broad uniform surfaces, are excellent "source regions" for the development of air masses. (An air mass is a huge body of air up to a thousand or more miles wide whose temperature and humidity properties are fairly homogeneous horizontally.) Air masses that developed over the Pleistocene ice sheets were cold, as are those that develop during our winters over the uniform snow-and-ice expanses of northern Canada and Eurasia and bring cold waves to surrounding regions when they move out.

We know what those air masses are like and how they behave. Air resting on the snow-covered surface becomes cold and therefore dense and heavy. It is clear and dry, and so allows heat gained during the day to escape into space rapidly at night. The air mass is stable, hard to lift and overturn, because its bottom layer is the coldest and densest, having been chilled most by contact with the snowy, icy undersurface. By itself it is not a good precipitation producer.

The atmosphere over a cold-surface source region such as this is a fairly permanent high-pressure center with air settling and moving outward from it. A cold air mass bursting out may move swiftly and bring sharp temperature drops. Its leading edge is a cold front. So powerful is this high-pressure air mass that it maintains its characteristics over long distances. We know how a winter outburst of cold continental air from snow-covered Canada, for example, can bring not only cold waves to northern United States but also severe freezes to the subtropical South. This "norther," still unchecked, though warming by then, can whip over the Gulf of Mexico in gales, and if strong enough can still blow with gusto across southern Mexico onto the Pacific Ocean.

We can imagine how much more intense the high-pressure system over a permanent ice sheet must have been and how much more dominating and far-reaching were the cold-carrying air masses which it was continually creating. That ice-cold source region existed year-round, not just in winter as is now the case; now the snowcover in Canada and northern Eurasia melts off in summer. And the ice-sheet source region was not down on ground level, but at a superior, colder elevation. From the roof of that frictionless, flat-topped dome, downdrafts of heavy air drained out on all sides. Cold waves were sent out frequently at all times of the year, blowing snow and cold beyond the margins of their stronghold, spewing their kind of habitat to outlying regions, readying them for capture.

But the ice sheet's cold blasts were not without opposition. The atmosphere strives for equilibrium, and warm air masses continued to approach and challenge the advancing ice and its missiles of frigid air. Important among the warm air masses were the maritime ones whose source regions were the large, uniform surfaces of warm oceans. They were opposites of the cold ice-sheet air masses. They were soaked with water vapor and were warmed at the base and therefore were lighter, unstable, and easily lifted. This same type of air mass is today's main bringer of precipitation to the northern continents. In North America tropical-maritime air from the Gulf of Mexico and the lower latitudes of the Atlantic Ocean is the principal supplier of rain and snow east of the Rockies, even to northern Canada. This air flows northward unobstructed up the low, open Mississippi River basin and over interior Canada into the subarctic. While there is no landform barrier now in the North American interior, just open plains and low hills, during glacial times a prominent barrier did stand there—the ice sheet, like an irregular plateau escarpment several thousand feet high. And barriers induce precipitation.

The warm, moist maritime air masses moved inland. Although light and buoyant compared to heavy cold ones, they constituted a front line of attack against the ice, and when they contacted it long enough they caused it to melt and weaken.

With misfortune often comes some benefit, and these same ruinous warm air masses carried the vital substance the ice needed to stay healthy and grow—moisture. If they were lifted and cooled to the saturation point they had to release this precious freight. And when they met the edge of the ice sheet they rose as though they had met a plateau or mountain range.

Ice sheets serve as landform barriers as surely as if they were constructed of rock. In fact, if we stretch a definition we could say glacial ice is a rock, one with an extremely low melting point. More specifically, we could say it is a sedimentary rock whose beginnings were not unlike those of sandstone, for example. Both are aggregates of settled mineral grains held tightly together. Compare snowflakes to grains of sand. They are both deposited and accumulate in layers, snow on land and sand in the sea. As snow changes to granular firn it is even more like sand. The grains in both cases are cemented together and compressed, the firn becoming compact ice and the sand becoming sandstone. Then if there is pressure and contortion, the structure changes and recrystallization takes place, and the sedimentary rock becomes a metamorphic rock. Just as granular sandstone may be metamorphosed, or changed, to harder crystalline quartzite, so relatively soft ice derived from firn recrystallizes into harder glacial ice. (The fact that glacial ice is plastic makes it no less like rock, for even "solid" rock becomes plastic when subjected to great underground pressure and changes shape or "flows.")

Over years of slow accumulation the ice sheets built themselves to plateau elevations so high that they covered all but the tallest mountains, up to heights where it was always freezing cold because of altitude alone. Like other high topographic features, they lifted and cooled the approaching air masses and drew from them the snow they needed to continue their conquest.

Precipitation was also induced by cold air masses dispatched by the ice sheets. They could do their work down-continent, far from the ice sheets, as well as near them. Continentally dry in origin, they could not themselves supply the snow required to keep the ice sheets strong and mobile, but they could wring snow from the subtropical and tropical maritime air masses coming from the warm oceans— in the following way:

A heavy, stable air mass, though invisible and moving, is as able as a stationary mountain range to lift another air mass. When two air masses collide the lighter one rises over the heavier one. This is what happens in cyclonic low-pressure storms, the "lows" we experience every few days in middle latitudes. Warmer air is actively rising over colder air and widespread cloud results, usually with precipitation. In pictures of Earth taken from satellites, which show the swirling, marbleized pattern of cloudy and clear areas, one can see the broad sweep of air masses and the interaction between contrasting ones.

Where masses of cold air from the snow-and-ice regions met masses of warm air they tangled in storms, not only near the ice sheet but also far down-continent. For hundreds of miles beyond the ice sheet in some places snow carpets were laid like welcome mats preparing the way for the entrance of the ice sheet into newly attached territory.

What blizzards and drifts there must have been then, such as we have never seen. Greater than those which occur now in Antarctica or Greenland, for in middle latitudes where the ice sheets were reaching, the air masses were much more copiously supplied with moisture.

But many of the storms were the nonviolent type, those dull, gray "lows" with which we are familiar. Often precipitation would come as rain, particularly in warm spells, which even an ice age has. Rain as a weapon of warmth could soften and melt the ice's surface. Still the ice froze some of the rain on and within itself. When rain fell or when warming caused surface melting, water seeped into loose firn and poorly consolidated ice, and then when the temperature dropped the water froze in the interstices, filling the hollows and strengthening the ice's inner bond.

As the ice sheets became larger they exerted more influence, causing latitudinal climatic zones to migrate equatorward as they do now in winter when they "follow the sun". Cold air masses took control of more territory. The middle-latitude belt of convergence where warm and cold air masses met moved equatorward too. The storms, or lows, along that belt swirled like slow eddies drifting from west to east. An occasional one would pass over the ice sheet, depositing snow there, and that was enough to nourish the interior, for there practically no melting occurred. Most of the snow fell near the margin of the ice sheet, helping it expand. The ice sheet's main zone of growth was its outer few hundred miles, and it grew most in a southerly direction. Although air masses from cold oceans contributed moisture to the ice sheets, they contributed less than those from warm oceans because they evaporated less water vapor, and held less, due to their lower temperature. And when the Arctic Ocean was ice-covered its water could not be evaporated and it was no better source of moisture than a land surface.

Stronger-than-usual temperature contrasts between warm and cold regions resulting from the ice sheets' presence may have increased the vigor of air-mass circulation and stepped up evaporation from oceans. So although oceans cooled somewhat as glaciation spread and evaporation from them should therefore have been less, increased atmospheric circulation probably kept evaporation high and helped distribute moisture.

We have seen how the ice sheets induced precipitation by cooling the warm, moist air masses that invaded their domain, but even when cooling of air masses could not produce precipitation it might produce condensation at least. When air is cooled to the dew point (saturation level) moisture is released. Cloud, fog, and other forms of condensation were valuable aids to the ice. From fog the ice could collect rime, which is like the white on freezer walls; and when surface condensation occurred in below-freezing temperatures, white hoarfrost ("dew" below 32° F.) would form on the ice surface. Ice deposited in these ways was only a modicum of nourishment compared to a good snowfall but it made its contribution. Fog and cloud served the ice in yet another way, by protecting it from the sun's direct rays and reducing melting. So the ice helped make its own shield.

But even when clear sky prevailed, as it did much of the time in the stable air over the ice sheets, the ice parried most of the sun's destructive rays. The ice sheets were immense mirrors that reflected insolation (solar radiation) back into space, absorbing only a small fraction of it. (Recall how you squint when you step outdoors on a sunny day after a fresh snowfall.) Sunlight that is reflected does not heat the surface. Incoming solar short-wave radiation must be absorbed at the surface to be converted to long-wave radiation (heat).

Fresh, clean snow has an albedo (ratio of amount of light reflected to that received) of over 90 per cent. Some snow measures as high as 98 per cent. Older, soiled snow and ice have lower albedos, but they may still be among the highest on the earth's surface. Tests made on tundra surfaces, like those that were near the ice sheets, showed that on clear days tundras register albedos of about 90 per cent when snow-covered, and only about 10 per cent when they are without snow.

Periodically the ice suffered from lack of snow and turned gray. When dust brought by winds dirtied its surface and thereby let more sunlight be absorbed, melting was facilitated. Then fresh snow would come and once again the ice in immaculate silver mail repelled the aggressor.

But the sun had more than one way to deliver its warmth. Even though its insolation was repelled by the high reflectivity of the ice sheets and snowy surfaces, it was being absorbed elsewhere in darker peripheral places. Radiation from those receptive surfaces warmed the air above and that heat could be transported by the atmosphere. Warm winds kept bombarding the ice, and not all could be blustered away. They could come and deliver warmth though thick clouds protected the ice sheets from direct sunlight. They could come even at night. Infiltrating up to the ice between outbursts of cold, they bathed it in warmth and weakened it.

Water too carried stored solar heat toward the ice. Small water bodies were no deterrent to the growing ice sheets, which just ate them up, but oceans were. Where an ice sheet reached the coast, waters lapped incessantly against it. Oscillation of the waves and the lifting and lowering of the tides broke off chunks of ice. Those icebergs counterattacked, chilling the water as they melted. Cooled water would drift away and be replaced by warmer water, but the ice sheet too had its reserves and still advanced on many ocean fronts.

Eventually the coldest adjacent ocean waters chilled to the freezing point, which is lower for salt water than for fresh water—about 27° F. (Ice, of course, can be much colder than 32°, that figure being its highest possible temperature.)

In quiet bays and inlets sea ice first formed. As the water became freezing cold it took on an opaque look. Frazil crystals formed—tiny needles and thin plates—and as they thickened the water became slushy. Then the surface froze solid, but wind and waves broke up that thin top layer and buffeted the pieces about, giving them round, lily-pad shapes. During a later cold spell they froze together once more and with time the ice cover became thicker and stronger and grew out from the shores.

All the allied defenses of land and air and sea could not restrain the cold conquistador. Its massive chilling presence alone refrigerated the whole area round about it. The more it grew the more it protected itself by this tremendous reservoir of cold.

Land beyond the ice sheets was subjugated by another ally, permafrost. How great an area was affected is not known but it must have been considerable. Permafrost must have extended hundreds of miles beyond the ice in the continental interiors, for even today it still grips our cold lands—nearly half of both the USSR and Canada (to south of Hudson Bay) and most of Alaska, plus other high-latitude and mountainous regions around the world.

Within the frozen ground were veins, wedges, and layers of solid, relatively clear ice. These grew larger with time because ice in the ground attracts water from the soil. As permafrost crept into an area the vegetation that needed normally deep soil and good drainage died or withdrew; and many animals left too because of the changing climate and changing food supply. The contagious cold was spreading, and wider areas were being conditioned for take-over.

But the ice sheets could be hurt. Though they were formidable and were gaining, the forces of warmth persisted. During summers and warm interludes warmth managed to wither a good part of the burgeoning ice and stunt its growth. But during winters and cold periods the proliferating sproutings of ice became sturdier and more firmly rooted and branched out in new directions.

The ice would expend great surges of energy and then there were times when its front lines would falter and drop back. The larger the ice armies became the more they needed reinforcements. And the reinforcements continued to come as zillions of snowflakes kept parachuting down, behind the lines and in front. The observant naturalist Thoreau said that fallen snowflakes look "like the wreck of chariot-wheels after a battle in the skies." In a way they were.

We have seen how the ice sheets controlled their environment, making it favorable for their growth, and engendered their own gigantism. How could self-sustaining monsters like these be stopped? Like blind, amoebic organisms they bulged out this direction and that, whichever way pressure nudged and snows beckoned. No obstruction hampered their progress. They captured one horizon and moved on to the next. Progress was at a creeping pace, but what was time? Something measured by a clock, or a calendar, or a season-counting intelligence? Something that did not exist. Their ultimate destination was to be the sea unless they melted beforehand, but they needed to travel no special direction and were bound for no place in particular.

No physical feature on Earth can stop a glacier's spread as long as reinforcements of ice keep coming to the front faster than warmth's melting and evaporation can destroy them.

And the ice sheets were huge now. The Scandinavian ice plateau may have mushroomed to a height of over a mile and a half, and

North America's to about two miles! These estimates are arrived at by considering the heights of mountains that were scraped over by the ice, by figuring slope angles of ice domes, and by judgment based on knowledge of Antarctica and Greenland ice sheets whose elevations reach about 12,000 and over 10,000 feet, respectively.

In some places the ice front was like a sloping wedge and in others like a shattering wall pushed from behind, tumbling its freshly cracked ice blocks ahead of itself.

If an ice sheet met hills its leading edge rode up the slopes, and it squeezed into the valleys and around the hills in a double envelopment. When the thicker ice came it flooded up and over the tops.

If it met mountains, the same pincers maneuver, only on a larger scale. Many mountains already had ice caps or small glaciers on them and these local militia aided in the capture of the region. Whole mountain areas were buried as the smooth-topped dome moved in. The highest of peaks here and there protruded above the sea of ice like islands. The Eskimo word for such a glacier-surrounded mountaintop is *nunatak*.

What if the ice sheet met a river? It would move across as if the river were not there. Where it first entered the river's course it would act as a dam. River water would pond up on the upstream side, creating a lake which would rise and spread until it found another outlet and spilled over there, perhaps into another river's watershed.

Over plains the ice moved with ease, obliterating old drainage patterns. Where forests stood in the way it snapped off the trunks and crunched right over the toppled trees.

Even a lake was no deterrent. It was merely another low spot to traverse, just a widening of a river behind some obstacle the river had not yet cut away. The ice waded in and kept on going.

But what happened when an ice sheet met the sea—the sea which has no end as land has and which makes up 71 per cent of the earth's surface? Could the ice sustain its advance against this water warmer than itself? It is amazing on how many fronts the ice did overpower the sea.

Floating ice shelves that jelled in quiet bays spread out to meet the stormy sea. Some shelves were just frozen surface water but others were glaciers extending themselves out beyond land. They were anchored at the shore and on islands and elevated parts of the ocean floor. Shelves that fringe Antarctica's coast now show what the moving ice shelves were like. Ross Ice Shelf, the largest, moves forward in its bay like a floating glacier. It is fed by glaciers behind it, which slide off the land, by snow it receives, and by sea water which freezes to its underside. This great shelf now covers 160,000 square miles, an area larger than California. No fragile film is it, but an over-the-sea extension of the continental ice sheet. Its outer edge rises cliff-like 200 feet above the lashing sea, breasting the roughest storms on Earth.

Ice has a unique characteristic that here comes into play. The general rule is that a substance contracts as it becomes colder. But water expands when it turns to ice. Therefore ice floats although it is colder than water. This is a critical factor because when it forms a solid sheet over the water it nullifies the water's ability to warm the air. It gives water the same continentality as a land surface and eliminates the moderating effect it has on climate. An advancing ice shelf can collect snow upon itself and feed itself just as a land-based glacier does. Its albedo is as high as any other snow-covered surface. Water can melt an ice shelf from below, but this loss can be replaced as water freezes to the shelf at other times. Eventually the shelf reaches a line beyond the headlands where storms and waves tear at its outer edge and icebergs calve off. Beyond it is a buffer zone of cold water filled with bergs and sea ice which freezes solid in the colder seasons and breaks up in pack ice in the warmer ones.

Where the ocean is shallow or where an ice sheet is heavy enough the glacier front may not be floated or broken off as it enters the water, but may instead grind right over the ocean floor as though it were on dry land. In that case it merely pushes the ocean back. Continental shelves were not difficult to wrest from the sea. They are the gently sloping, underwater extensions of continents, at most 500 or 600 feet under water. The ice sheets in some places barely got their feet wet as they entered the sea.

Even valley glaciers streaming out of fiords had success against the sea. Since ice can cut below base level, glaciers could deepen their valleys well below sea level even before they reached the coast and continue cutting troughs in the ocean floor a considerable distance from shore before they were buoyed up or broken off. Later, when the ice was gone, the ocean entered these inlets, penetrating long distances up the old valleys. Fiords are noted for their great depth and their high, protective walls. Through them ocean ships requiring deep channels can travel far inland. Sogne Fiord of Norway is 4,078 feet deep. That is about four fifths of a mile! It extends inland 115 miles. And Messier Channel in Chile is 4,167 feet deep.

Offshore islands were captured and often overridden by the ice sheets. As water passages between islands and the mainland were closed by ice shelves or solid glacial ice, currents were blocked and the sea's moderating effect was lessened.

As ice piled up on the land the oceans shrank, for moisture evaporated from the oceans was dropped on land and imprisoned there. In its frozen state it could not drain back to replenish the oceans. Much of the continental shelf area was exposed and shallow parts of the ocean became dry land. Many islands were then tied to the shore and some of the large land bodies were connected.

How far did sea level drop? That is not the easiest mathematical problem. One needs to know how much ice was on land, how much water that would equal, what the ocean area was, and a few other basic facts. The problem is further complicated. Ocean bottoms rise as water weight above them is reduced. Continents sink under the weight of ice sheets, but not evenly. They tilt down on the heavy side, and they do that gradually. Also, continents and ocean basins are always rising a little here and sinking a little there, with or without ice sheets being on the scene. Old shorelines must be used carefully as registry lines, therefore, but they help show where the former high-water level was. The low-level shorelines, of course, are under water, for oceans have been rising as the ice has been melting. According to present calculations, as glaciers grew to maximum size sea level dropped something like 450 feet, to a level about 300 feet lower than now.

On the glaciers rolled, scoring victory after victory, ravishing the land and forcing even the sea to yield. If anyone had been watching this conquest they would have had to wonder how far it could go.

Space enthusiasts find it easy to believe that if inhabitants of another planet had attained a high civilization earlier than we and had been surveying other worlds with telescopes and spaceships, they might have observed our ancient ice sheets. Perhaps they watch our present ones. The thought is not utterly absurd, for our own astronomers, even amateurs with small eyepieces, observe the seasonal formation of white caps on Mars.

Imagine hypothetical space observers through long, sidereal lengths of time watching Earth's ever-changing aspect, her land and water bodies assuming different shapes, and then—strange white spots began polka-dotting the surface. A subsequent check a long time later would report that some of those spots had grown considerably. The largest white areas would have widened like buds bursting into flower. New spots popped up and they too expanded. Adjacent ones merged into large glittering crusts. Those were visually augmented by sheets of snow which lay over much of the globe, especially in Northern Hemisphere winters, and by the white coating of ice on the oceans. The white brilliance seemed on its way to encasing the whole world. But no. Over many more long years the white withered away. However, it re-formed and grew again, and then withered away once more. Several times that happened.

No one knows what caused the ice to form, nor why it goes away and comes back again. We have theories aplenty. But the real cause of the Ice Age may never be learned on this planet, particularly if the Ice Age should have been caused by conditions outside Earth about which we have no knowledge. Perhaps we shall find the answer to that mystery recorded on another planet, particularly if the tireless geologic secretaries and accountants there have drudgingly recorded natural events in rocks and sediments as they do here. In some distant world may lie the Rosetta Stone that will enable us to decipher and translate our own cryptic, geographic record and so tell things about this secretive world and ourselves that we, so close, cannot perceive.

CHAPTER 7

THE ELUSIVE REASON WHY

He does great things which we cannot comprehend.
For to the snow he says, "Fall on the earth" . . .

By the breath of God ice is given,
    and the broad waters are frozen fast. . . .

Whether for correction, or for his land,
    or for love, he causes it to happen.

ELIHU (to Job)

IF ALL our glaciers were gone and no knowledge of them
remained, could we in our wildest dreamings think such things had
been?

Or even if just a few valley glaciers existed in faraway places, but
all large glaciers including those of Antarctica and Greenland disappeared, who of us would conceive of ice sheets a mile or two high
overrunning continents?

Therefore, early scientists of Europe ought not be disparaged for
their failure to deduce until the last century that glaciers of tremendous proportions had been mounded over the very land on which
they lived, even though that land had been long inhabited and
known in detail. What a revelation it must have been when they began insisting on objective, not superstitious, answers to questions;
dared explore the byways of the unknown; and confronted the Ice
Age apparition coming into focus out of the blank past like a gigantic
iceberg looming out of a midnight fog.

People had long wondered about the source of stony drift and erratic boulders that were spread over northern Europe. Southern Eu-

rope and lands beyond did not have them. Various explanations sufficed. That unnatural coating of "boulder clay" and erratics supposedly resulted from earthquakes or Earth's encounter with a celestial body; or the Alps had risen suddenly with violent explosions that scattered the drift and boulders about. Later the Biblical Flood of Noah's time became the approved explanation. The floodwaters, it was taught, had "drifted" the jumble of rocks, clay, and sand over the land. So it was that those strange deposits came to be known as "drift." Only after perceptive mountain climbers and inquiring scholars and naturalists who came close to valley glaciers listened seriously to mountain peasants' stories of what their glaciers did, and took notice of how those glaciers moved and transported mixed rock debris which resembled drift in the lowlands—only then did the concept of the Ice Age begin to materialize in imaginative minds.

No, we should not scoff at the early scholars' slowness in solving the riddle of the drift and envisioning a continental ice sheet, especially when Antarctica had not yet been discovered and Greenland's ice cap was still unseen except for its outer edge. Here we are, several generations advanced, with a wealth of scientific tools and the ability to explore anywhere on Earth and the moon, and still we have not been able to explain *why* the Ice Age occurred. There must be a hundred theories but they all have "holes" or lack substantiation, and none can be proved. Not one commands anything like general acceptance.

A satisfactory theory of the cause of the Ice Age should clarify several points: (1) why ice sheets formed in the first place; (2) why they withdrew; and (3) what caused the oscillations of the ice, that is, the repeated advances and withdrawals—the separation of the several glacial stages by warm interglacials, and the smaller fluctuations of the ice within each glacial stage.

Theories range from simple to complex. The simplest are pure supposition. One of these, the "cloud in space" theory, suggests that our solar system passed through a cloud of interstellar matter which diminished the heat that planets received from the sun. This idea has been carried a step further to say that the myriads of cosmic dust particles left in our atmosphere by such a cloud would have served as extra condensation nuclei, those tiny particles on which condensation forms. Therefore with more nuclei there could have been more condensation and more precipitation—more snow for glaciers in colder regions and more rain in warmer ones. Then, as the screen of nuclei settled out, the climate would have become warmer and drier again.

The "variable sun" theory is still simpler. It even omits the spatial cloud and proposes merely that the sun at times emitted less heat and that at those cooler times glaciation occurred. Solar radiant energy does vary in amount, but rarely more than 2 per cent. It varies also in quality, and it is thought the relative amount of ultraviolet light emitted might have been significant. We shall soon be learning more about present variations in the sun's output as satellites orbiting in clear space accumulate reliable solar-radiation data.

Whether cosmic dust clouds or variability of solar radiation caused the Ice Age may not be known until other planets of our solar system are explored and it can be seen whether nature imprinted records of past colder periods on them also and whether those cold periods were synchronous with ours. What throws doubt on the variable-sun theory is that, as far as we can tell, our sun through all of historical time has been characterized more by constancy than by changeableness. Yet one cannot say what happened in the prehistoric past or what could happen tomorrow.

In other extraterrestrial theories glaciation has been attributed to Earth's passing through "cold spots" in space; or to "tides" in the atmosphere produced by the moon's attraction; or to Earth's capturing another planet which revolved about us displacing the poles, moving the crust and creating tidal floods; or to unusual behavior of comets, asteroids, and meteors; or to variable radiation of the Milky Way.

An "astronomical" theory sees the Ice Age caused by known, measurable irregularities in Earth's position relative to the sun which influence the distribution of solar heat over the planet. These irregularities, which are real, are in the (1) eccentricity of Earth's orbit, which affects average temperatures over the whole world; (2) inclination of Earth's axis to the plane of its orbit (the present value is 23° 27′ but over a long-term cycle it tips between 21° 58′ and 24° 36′, affecting seasons); and (3) wandering of the perihelion, the point of the orbit closest to the sun. (It is interesting to note that at present Earth comes nearest the sun in the Northern Hemisphere winter, in early January.) These irregularities go through plottable cycles of 91,800 years, 40,400 years, and 20,700 years, respectively. Interaction of the three cycles may periodically have cooled certain segments of the planet enough to give rise to glaciation, according to this theory; but its main flaw is that it requires that the glacial stages alternated between the Northern and Southern hemispheres, one hemisphere being warm while the other was being glaciated, and such is not known to have been the case. It does not explain either the many irregularly timed small-scale oscillations within the larger stages. Surface temperatures for different latitudes have been computed backward in time for thousands of years on the basis of these cycles but results are inconclusive.

The most challenging theories are not the extraterrestrial ones but those that put the cause of the Ice Age right on Earth, because the chance of proving or disproving them is within the realm of possibility without waiting to put geologists on other planets.

In the "volcanic" theory Earth supposedly cooled in response to sky-darkening volcanic eruptions. The eruption of even one large volcano can darken the sky appreciably, and it is possible that the eruption of many could have altered climate, if only temporarily. When a volcano erupts it can shoot great volumes of gas, ash, and fine dust high into the air. When the dust particles are carried to the upper atmosphere they may float there like a veil for a number of years, scattering and absorbing incoming solar energy and reflecting it back into space. Such a veil can lower temperatures at the earth's surface. According to this theory also, when the particles are in the lower atmosphere where weather phenomena occur they may serve as nuclei for condensation droplets and crystals, helping clouds to form and under certain conditions increasing precipitation.

In 1815 Mount Tambora east of Java blew up about thirty cubic miles of rock, which equals about 185 Vesuvius-size mountains, and the region around it was in complete darkness for three days. The dusty discharge joined that expelled by a Philippine volcano the previous year, and both dust clouds drifted about the world in the upper atmosphere. Their screening out of insolation could have been the reason that 1816 became famous as "the year without a summer," the year that New Englanders called Eighteen-Hundred-and-Froze-to-Death.

Following the 1883 eruption of Krakatoa, an island volcano in Indonesia, the solar observatory at Montpellier, France, recorded radiation 10 per cent below normal for three years. Two thirds of the island was torn away, leaving 1,000 feet of water where land had been. Three hundred Indonesian towns were destroyed by that blast, which was heard 2,200 miles away in Australia and 3,000 miles away at Rodriguez Island in the Indian Ocean. Sunsets around the world were exceptionally red for two years after that.

Heavy, sustained volcanic activity, some contend, would have brought on glaciation, and interglacial periods would have occurred when activity subsided and skies cleared. It is true that mountain building with concomitant volcanism did precede and continue into the Pleistocene. But the theory weakens when asked to explain why ice ages did not occur at times in the past when volcanism was even greater. That volcanic dust can screen out the sun is granted, but it is hard to believe that volcanic eruptions could have been numerous enough and continuous enough to maintain coolness for the thousands of years needed to build large glaciers. If volcanic dust was abundant enough in the atmosphere to have caused the Ice Age, then there should have been copious, widespread fallout of this dust during each cold stage, but such fallout has not been found in Earth's sediments. Some has been found deep in the Antarctic ice sheet though. Doubt is cast upon the volcanic theory by those who feel, contrariwise, that lingering dust clouds would have hampered glacier growth, instead of fostering it, by acting as a heat-holding blanket, warming the earth instead of cooling it, and that fallout settling on glaciers around the world would have grayed them and so aided their melting instead of their growth. Without being able to accurately date Pleistocene glacial stages and times of exceptional volcanic activity, and so correlate them, it is impossible to verify the relationship.

Another school of thought is that the decrease of certain gases in the atmosphere caused the lowering of temperatures. Water vapor and carbon dioxide are particularly suspect in this regard. Both of those gases allow solar short-wave radiation to pass through freely, but when that incoming radiation has been absorbed at the earth's surface and changed to heat, the outgoing long-wave radiation is readily absorbed by them. What these gases do is illustrated by what windows of greenhouses or closed automobiles do. They let direct sunlight through but they trap heat that results and keep it from escaping. You have noticed how nighttime temperature drops more slowly in humid weather than in dry weather because there is less water vapor in dry air to absorb heat and retard its escape into space. Carbon dioxide performs a similar function, and so even though it constitutes only 0.03 per cent of the volume of the atmosphere, it is critical in the regulation of temperature. It is said that if its volume were reduced to half its present amount the earth's average surface temperature would drop about 7° F. (more than enough to call back the Ice Age). The amount of carbon dioxide in the atmosphere does fluctuate because oceans are capable of absorbing and releasing it. Cold water can hold more of it than warm. And plants and rocks contain it.

If carbon dioxide, water vapor, and other gases were not responsible for starting the Ice Age they may have had some effect on the warm-and-cold variations within it. It is conceivable that their amounts in the atmosphere might have fluctuated considerably when oceans were alternately increasing and decreasing in size and cooling and warming; when continents were undergoing transformation; when the vegetation cover was changing; and as ice sheets covered and re-exposed up to 30 per cent of the land surface.

The atmosphere—its composition and circulation—figures in a number of Ice Age theories, but it is not the sole distributor of the world's heat. Oceans, through their great currents like the Gulf Stream, transport possibly up to a third as much heat from low to high latitudes as does the atmosphere. They carry equatorial warmth to polar regions as warm currents and then swing back as cold currents flowing equatorward to absorb a new supply of heat. And there are smaller, more regional currents. When land masses were being connected or separated, when the submarine topography of ocean basins was altered, when sea level rose and fell, then some currents were redirected and the climates of regions affected by them changed. But how great this readjustment was is not known. Maybe redirection of ocean currents did make snow accumulation possible in certain areas where it had not been possible before. It is important whether air masses determining climate of a given region have passed over warm water or cold, because the temperature of the water affects the temperature, moisture content, stability, and energy of those air masses. But it is not natural for ocean currents to shift back and forth from old to new circulation patterns with each glacial and interglacial stage, requiring that certain land areas or submarine barriers time and again sank from and rose to the same positions to accomplish this.

Most theories of the Ice Age try to show how it occurred because either (1) the temperature dropped or (2) precipitation increased, or both. There are striking exceptions, however.

One jolting theory is that the Ice Age was brought on not by falling temperatures but by rising ones! The reasoning, which is not totally illogical, is that moderately higher temperatures would increase both evaporation from oceans and atmospheric circulation and this would bring heavier precipitation to high latitudes. Also there would then be more cloud cover, which would protect accumulated snow and ice against melting. A tremendous amount of heat energy was required to lift from the oceans all the moisture that built the ice sheets. To get the higher temperatures the theory falls back on the good old sun and has it increasing its energy output at the required times.

Contradicting this theory is evidence that cooling occurred gradually during the long Tertiary Period preceding the Ice Age and then more abruptly during the several glacial stages. This evidence is seen in the soils, plant pollen, marine-life remains in ocean sediments, and other fossils left from those times. Hawaii and New Guinea in the tropical Pacific grew glaciers during the Ice Age and so did many inland mountains in equatorial latitudes. That greater warmth should have been required to produce these expressions of coldness in already tropical, moist regions seems an awkward, unnecessary distortion of geography. Warm air does have greater evaporation and moisture-holding capability, but a combination of glacier growth and warmer temperatures strikes a discordant chord of incompatibility. Some climatologists believe that precipitation during glacial maxima could have been even less than now (dispensing with the need for higher temperatures). It is not necessarily the *amount* of snow that causes ice sheets but the fact that snow, however little, lasts. The Antarctic ice dome today, in terms of precipitation received, is the driest large area on Earth, drier overall than any large desert.

Still, to some ways of thinking, *cold* and glacier growth are antagonistic. It is argued that it is not climatically proper to have increased precipitation and lowered temperatures existing together. Northerners know the expression "It's too cold to snow." Usually during snowstorms the temperature is above average for the time of year, and severe cold comes with the clear air after the snowy weather has moved on. Heavy precipitation is more generally allied with warm or high-sun seasons and with warm regions than with cold, though there are exceptions. In cold regions it just appears that winter brings more precipitation than summer because piled-up drifts and plowings of many snowfalls stay around a long time whereas the heavy downpours and drenching rains of spring and summer are quickly gone and forgotten. On the average about twelve inches of snow equals one inch of rain. Assuming that precipitation was substantially increased over large parts of the earth to form the glaciers and ice sheets, then, in this theory, over other large areas evaporation should have substantially increased also, and therefore at least in those evaporation areas it is supposed that climate warmed. Did some parts of the world warm while others cooled? Did increased warmth prompt glaciers to begin and then, after that, did the temperature drop? Or were even small amounts of snow (which did not melt away) enough over the centuries to create glaciers?

There we bump into one of the toughest questions: which came first to start a glacial stage, cooling or increased snowfalls? Did cooling take place first, with the result that more of the precipitation fell as snow instead of rain, and that more of the snow that fell was preserved? Or did precipitation—that is, snow—increase first (even perhaps with relatively high temperatures) and then did the ice cover which resulted cool the land, air, and sea?

Imagine the confusing task of the paleoclimatologist. It is hard enough classifying and mapping today's multiple types of climate and tracing the three-dimensional circulation of the ever-restless atmosphere and oceans. But complex as that job is, how much more complicated it is to reconstruct climates as they might have been in times past. The susceptibility of various sites to glaciation would have been determined not by just a few simple conditions but by a whole interrelated set of them—latitude, elevation, clarity of the atmosphere, position in relation to storm tracks, water bodies, ocean currents, prevailing wind direction, and other factors, including the possibility that the poles may have had different positions or that the continents may have moved.

When solid ice covered the upper middle latitudes the transitional zones between frigid and warm climates were telescoped closer than now, so there were steeper temperature and pressure gradients, sharper contrasts between warm and cold zones. When frigid zones lay only about half as far from the tropics as they do today, what clashes of cold and warm air masses there must have been, what vigorous storms, what sudden and pronounced weather changes! What unguessed weather wizardry could then have been brewed from the turbulent mixtures in the atmosphere's big, round cauldron. And in the dark ocean basins what conflicts and mixings took place as frigid waters spilled from ice sheets, sank to the bottom of the oceans, and spread outward.

What caused the Ice Age?

Some say the amount of heat emanating from Earth's interior lessened and has fluctuated since.

Some theorists point a finger at Antarctica as the regulator of worldwide glaciation and deglaciation. As things stand now, that ice-capped continent is likely to have been one of the first spots on Earth to grow Pleistocene glaciers, and it might well have retained a large part of its ice cover through interglacials as it is doing now. This sequence of events has been postulated in one theory. During interglacial periods like the present, ice accumulates on Antarctica, building up higher and higher. While the top remains cold the base of the ice sheet is warmed by heat from Earth's interior and by pressure from the ice above, which keeps increasing as the ice thickens. Finally these combined warming effects melt the ice at the bottom, freeing it from underlying rock, and the water at the ice's base acts as lubrication causing the ice to flow faster. Friction on rock produces more subglacial heat and more melting at the base, helping the ice move even faster. As the outward-flowing ice reaches the surrounding oceans it supposedly spreads over them as floating ice shelves. The encircling ice shelves theoretically would have covered ten million square miles of ocean and that white surface would have reflected so much sunlight away from Earth that the planet's absorption of it would be reduced by 4 per cent. Some say the ice would have come off the continent not as a shelf but as icebergs. Either way, the oceans would have been chilled, cooling climates around the world. Then glaciers would start to grow in the Northern Hemisphere as well; and the world would become still colder and glaciers would grow still larger. Then the pendulum would swing the other way. As ice slid off Antarctica the remaining ice sheet there would be thinner and less heavy, so the contact zone between it and bedrock would become colder again and the slippery water layer there would freeze once more, holding the remaining cover of ice to bedrock, and so the delivery of ice to the oceans would slacken and the ice build-up on Antarctica would begin again. This would be the start of an interglacial when ice stayed on Antarctica and the rest of the world was permitted to warm.

If we believe this theory we see ourselves now heading for another glacial stage, and regard the ice shelves and frozen seas around Antarctica as only precursors of much larger ice shelves to come. The volume of ice on Antarctica now is immense, making that the highest of the continents with a mean elevation of about 6,000 feet. In some places the ice sheet is over two miles thick. If that ice did spread out over the seas it would be a gigantic ice mirror all right. It has been calculated that if Antarctica's approximately 6,000,000 cubic miles of 'ice should completely melt, the level of the oceans all over the world would rise about 200 feet. In light of this it is of interest to note what the U. S. Army Cold Regions Research and Engineering Laboratory found at the rock base of Antarctica's ice sheet when its great drill bored down more than a mile through the ice. At the bottom is a thin layer of water!

Another finger points the opposite direction, to the Arctic Ocean, as the locale critical in the regulation of glacial and interglacial stages. One much-modified theory seeks to explain the climatic oscillations by supposing that the surface of the Arctic Ocean repeatedly froze over and melted and froze over again. Theoretically, when that polar ocean was open its ice-free surface was a large, wet source region providing evaporation for the heavy snow which helped glaciers become established on lands around it, including those largest ones on Greenland, northern Canada, and Scandinavia. As the ice sheets expanded they cooled the atmosphere and seas, sea level dropped as ice stacked up on land, and gradually the Arctic Ocean, which was becoming shallower and chilled by cold air above and additions of ice water and icebergs, froze over. Then the moist source region supplying snow no longer existed. But by that time, the theorists say, the ice sheets were large enough and had expanded southward far enough to catch precipitation from oceans to the south. Their growth stopped when they reached latitudes too warm for their survival. All this time other oceans besides the Arctic were cooling too, evaporation lessened as temperature of water and air dropped, and so also did the snowfall on the ice sheets lessen. The ice stopped growing and then retreated. An interglacial stage began. With melting, sea level rose, the seas and air gradually warmed as the refrigerating ice sheets shrank, and in time the ice cover of the Arctic Ocean was gone. Then the cycle reversed and once again, with that open moisture source available, evaporation and increased snowfall in circumpolar lands began the next glacial stage.

Like other theories presented, this "open and closed Arctic Ocean" theory is more elaborate than my short summary, and it also has its weaknesses. In response to criticisms its proponents have radically altered, even eliminated, some of the theory's original main points until the reasoning and sequence of events are quite different from those first offered. The defensive modifications this theory has gone through raise doubts about its soundness, but on the other hand it may be a stronger theory as a result of its having accommodated criticisms. It still has inherent shortcomings though.

It concentrates on the Northern Hemisphere and would have the changing conditions of the land-ringed Arctic Ocean generate warm and cold cycles for the whole world. Then, too, although the Arctic Ocean is made all-important in the initiating of the ice sheets much of the land around it was scantily glaciated. Large stretches of Arctic Siberia, some islands of northern Canada that lie right in that ocean, as well as much of Alaska, had only thin glaciers or were never glaciated at all. Of course, nearness to water is no guarantee of precipitation. We have deserts today that border oceans. But still, if the Arctic Ocean were the main causative factor in glaciation it would seem that its land rim should have had more and larger glaciers than it did. One can come up with reasons why it might not have, but that is laboring the point. On the basis of where the big glaciers used to lie one might consider the North Atlantic, not the Arctic Ocean, as the main moisture center for Northern Hemisphere glaciation. Most climatologists have felt that the Arctic Ocean was too cold to provide abundant evaporation and that it was only a secondary supplier of moisture for glaciers on surrounding continents.

Despite its weaknesses this theory has many supporters, and it continues to be studied carefully because the ice cover on the Arctic Ocean is diminishing and an open ocean is predicted. The climate of the Arctic Ocean shores will change if that happens. How and how much are the big questions. Some researchers have data that suggest that the Arctic Ocean may have been frozen throughout the Ice Age, through glacial and interglacial times, but this is still uncertain.

A broader "ocean control" theory contends that the key to glacial cycles was the fluctuating temperature of all oceans, not just the Arctic. Before the Ice Age, oceans were warm. Then, according to this theory, the land rose enough to make cool areas where glaciers could develop. Icy rivers draining from the glaciers, ice melting in the ocean, plus the chilled air gradually cooled the oceans. Oceans lag behind the land in their thermal responses, so this cooling process took thousands of years. Evaporation was less from the cooled oceans, so precipitation over most of the earth declined. Glaciers therefore wasted away and many lakes dried up. Then the land could have received and held more heat, the albedo effects diminished, cloud cover lessened, and rivers emptied warm water into the oceans. The oceans slowly warmed until again, after thousands of years, evaporation and precipitation increased, allowing glaciers to re-form. And so the cycles recurred. At present, the theory goes, oceans are relatively cool from melted ice, lakes are low, and warming glaciers are wasting; and when oceans become warmer glaciers will grow.

Oceans cover nearly three fourths of the planet; have abysses deeper than the highest mountains are tall; are greater in volume than all the land above sea level; have tremendous capability to store and deliver heat. They could, indeed, hold the key to the Ice Age.

In textbooks many of these theories have persons' names attached to them as their formulators. But names have been omitted here because most theories are not original with one person or even a couple of people. As you have noticed, some of the same ideas are incorporated in various theories in different ways. Every few years a "new" theory is proposed, but it is generally a revival or modification of an older one. It is given headlines in magazines and newspapers and then, after the critics have dissected it in the less-publicized professional meetings and journals, the public hears no more about it. Then another new idea pops up and is in the limelight awhile. When you hear of a startling new Ice Age theory it is best not to accept it too readily but to wait and see how well it stands up. Names are omitted also because the critics of a theory—who usually receive no mention—are often as important as the originator to a theory's ultimate form. By pointing out its weaknesses they help the originator revamp it and make it more plausible.

By now you have probably noticed the one great weakness inherent in many of the theories. They carefully explain the shift back and forth from glacial to interglacial conditions, but cannot explain what set the initial glacial cycle going. Because this is a flaw in so many theories it was left to be discussed separately here.

One widely accepted reason for the start of the first Pleistocene glacial cycle has been that the land was rising to higher elevations at, and shortly before, that time. With greater altitude comes increased cold, and increased snow on mountains. With larger land masses comes greater continentality, greater temperature extremes.

If you start with that premise, that rising land was responsible for starting glacial conditions, then you still need to say what caused the warm and cold cycles within the Ice Age. Certainly the land did not go up and down. The most easily accepted explanation is that energy from the sun fluctuated. A combination of the fluctuating-sun and rising-land concepts gives us the "solar-topographic" theory, one that has had many adherents. Its popularity has been due partly to its simplicity and reasonableness. Scholars find it safest over the years to line up with a theory that is not overly imaginative or complicated and does not strain for credibility or shake up the old planet too much. But this theory is no nearer to being proved than any other.

There are other explanations for the start of the first glacial cycle that do shake up the planet. One is that Earth had a collision with a giant meteor which changed the tilt of its axis and got things going. Another, which is taken more seriously, is that our planet is like an unsteady gyroscope and slowly wobbles on its axis; the wandering of the poles brought glacier-susceptible areas into cold high latitudes at some times, and at other times the poles were in such locations that land areas were not where they could grow glaciers. The poles do wander slowly even now and seem to have wandered more in the past. The magnetic "imprint" in many ancient rocks of different ages and locations is not aligned with Earth's present magnetic field. When those rocks formed, the magnetic poles were in places other than where they are now, sometimes thousands of miles away from their present location. Either the rocks—that is, the land masses— moved, or the poles moved, since the rocks formed. Various hypothetical paths of wandering poles have been charted using the complicated distribution of magnetic rocks as guideposts. This promises to be an absorbing and perplexing game among geologists for years to come.

The explanation of how the land masses moved in reference to the poles used to be that Earth's crust slipped about like a loose covering on a ball, eventually bringing about the right arrangement of land and water on Earth's grid to cause the Ice Age. Warm and cold cycles within the Ice Age were accounted for by later slippages.

Another theory involving moving land masses is that of "continental drift," which drifted in and out of favor for many years. It is now at a new crest of popularity in updated form known as "sea-floor spreading" and "plate tectonics," and may bring us some of the answers we are seeking. Before we examine it we need some background information.

This book is concerned with just Pleistocene glaciation, but in searching for its cause we need to look far back in time. There seem to have been earlier, ancient glaciations hundreds of millions of years before the Pleistocene. If so, then any valid glacial theory should explain not only our Pleistocene Ice Age but earlier ones as well.

The belief that ancient, pre-Pleistocene glaciations occurred is based upon *tillites*, deposits that look like old, hardened boulder-clay drift containing unsorted rocks many of which have facets and striations characteristic of glacier-dragged rocks; and in places the tillites lie directly over striated and polished bedrock that has the appearance of being ice-scoured. If these tillites are indeed glacially made, then there were at least two major periods of widespread glaciation before our "modern" Pleistocene Ice Age. One would have occurred at the close of the Paleozoic and one at the close of the Precambrian, and there may have been still others. (See Geologic Time Scale, page 11.) Those pre-Pleistocene glaciations occurred in some unlikely places, not only in cool regions like Greenland, Scandinavia, and Antarctica, but also in Africa, Australia, India, and eastern South America, all of which are warm today.

Furthermore, about the time old ice sheets are supposed to have covered areas that are now tropical, coral reefs were growing in some waters that are now subpolar, and coal beds—the remains of luxuriant warm-climate vegetation—were forming in now-cold places like Antarctica, Greenland, Spitsbergen, and Alaska. One can't help suspecting that something has been moving. Maybe the poles. Maybe the continents. Maybe both.

When you had your first geography lesson in grade school and looked at the wall map your eyes made puzzle pieces of the continents and fit the curves of the Americas into the curves of Europe and Africa. South America and Africa especially could snuggle close together. And then you forgot the obvious as your study was directed to geography of the present.

One can still play with that puzzle. Perhaps long ago, when ancient, pre-Pleistocene glaciation is said to have occurred on separated continents of the Southern Hemisphere, those continents were united in one land mass, a hypothetical southern continent that geologists call Gondwanaland, named for a region in India. On this supercontinent those areas of old tillites were joined together where one great ice sheet could have covered several or all of them. Presumably that area was then nearer the pole. Later Gondwanaland supposedly split apart into separate continents which moved to where they are now. Another supercontinent was supposed to have been Eurasia and North America combined before they split apart.

(If geologists millions of years from now were to discover scattered traces of Pleistocene drift and ice scouring in northern North America, Europe, Greenland, Iceland, and Siberia, might they conclude— as the envisioners of Gondwanaland did—that those regions must have been united at the time of glaciation and been located nearer the North Pole?)

Much support for the concept of moving continents comes from ancient rock formations and fossils. Rock formations ending at the coast of North America seem to match up in location and structure with similar formations in western Europe, and rock formations of South America likewise match up with counterparts in Africa, and there are apparent matches among the other Southern Hemisphere continents too. It seems as though the formations split apart after they formed. The break-up is believed to have occurred during the Mesozoic, the age of dinosaurs.

Recent discoveries in Antarctica of fossil bones of land reptiles and fresh-water amphibians that lived in other Southern Hemisphere continents some 200 million and more years ago are cited as proof that Antarctica was once united with those continents because these animals could not have crossed the ocean that now separates the land masses. The remains of ancient warm-climate vegetation in Antarctica give further support to the belief that it was once in a more northerly location. So now we have the picture of Antarctica splitting off from Gondwanaland and "drifting" south toward its polar position during the Mesozoic and perhaps Early Tertiary, and glaciation beginning as it did so.

Proponents of continental drift used to talk of continents sliding slowly across the ocean bottoms. Now they talk in terms of ocean floors spreading, carrying the split-apart continents away from each other as though they were on conveyor belts going in opposite directions. The rifts may have begun like the volcanic Rift Valley of eastern Africa or the Red Sea and slowly widened into ocean basins. Sections of continents and ocean floors are also seen as "plates" moving about, separating, bumping into and sliding against one another, causing mountains to form from the collision. So land masses may have joined as well as split. For instance, India may have moved north and pushed into Asia causing the Himalayas to buckle up.

The most convincing evidence of moving continents was found through seismic and magnetic surveying of the Mid-Atlantic Ridge, a curving submarine mountain range running from Iceland south along the center of the Atlantic Ocean floor. Sensitive instruments have recorded imprints of paleomagnetism in successive lavas that welled up and filled fractures along that ridge. The magnetic imprints in the rocks show an age pattern that is the same on both sides of the ridge: the youngest rocks are right next to the ridge running parallel to it on both sides; and progressively older rocks lie farther from the ridge, both east and west of it, in a mirrored pattern. The relative, matching ages of rocks on either side of the ridge are determinable due to the fact that Earth's magnetic field has reversed itself many times in the past. That is, the North Pole became the South Pole magnetically, and then it changed back. As the switch occurred from time to time, the magnetic imprints recorded the reversals and thus registered time intervals. (Reversals of Earth's magnetic field are considered a possible cause of mass extinctions of certain types of animals throughout the planet's history, including the Ice Age, and may have had an important bearing on other Ice Age events.)

Lava escapes to Earth's surface along the volcanic Mid-Atlantic Ridge and builds new ocean floor, which apparently moves away from the ridge to the east and to the west, carrying the continents farther and farther apart at an imperceptibly slow rate. If they moved, say, an inch a year, they would have traveled less than sixteen miles in a million years.

Ridges that behave the same as the Mid-Atlantic exist in other oceans, and somewhere—perhaps in the deep sea-bottom trenches— the ocean floors must sink back into Earth or else the world would be expanding like an inflating balloon.

Knowledge of sea-floor spreading is still in a confused state, but evidence in support of it continues to mount. Since this concept was realized geology has entered a new age of discovery as eye-opening and astonishing as that era when evolution and the Ice Age were first conceived of. Sea-floor spreading may in time supply an ex- planation for the more ancient glaciations and for certain regions having had different climate than they now have, but it is not a sat- isfying answer for the Pleistocene Ice Age. Not yet.

One can visualize Antarctica slowly "drifting" south to its polar location at the proper time, but the Northern Hemisphere land masses which became heavily glaciated appear to have moved mainly east and west from the ridge that is now in the central Atlantic, whereas the more critical direction climatically would have been north. Maybe they did move north too, allowing glaciation to begin, but as yet there is no sign that they moved any appreciable distance in that direction at the required time. Also, this theory fails to ex- plain the oscillations of the ice. By no stretch of the imagination can one believe that drifting continents could have backtracked for each interglacial period and then returned to their colder positions for the next glacial period and done this several times, causing the ice to advance and withdraw. Other factors must be involved.

By mixing continent drifting and pole wandering you can have infinitely complicated things going on in the past. Data are too sketchy to clarify the situation, so numerous versions of what might have happened are heard. Instead of the history of climatic changes growing clearer, it is becoming more complicated all the time.

Events that brought on our Ice Age may have been as amazing as the ice sheets themselves. Or they may have been surprisingly sim- ple. Even if fantastic things did happen to initiate the Ice Age it is generally agreed that by Late Tertiary—certainly by the start of the Pleistocene—the continents and poles were essentially where they are now. At least that is what we have to assume on the basis of what we know. And if the continents were drifting their rate of movement was too slow apparently to have been responsible for the marked geographical changes that occurred during the Ice Age.

Through most of the Tertiary and through glacial and interglacial periods climatic and vegetation zones seem to have remained generally concentric about the present poles, indicating that there was no significant latitudinal change in the location of continents. Tertiary corals which required continuously warm water are distributed symmetrically around the present equator. During the Ice Age, positions and fluctuations of snowlines on mountains at all latitudes responded, on an average, in a way that shows the poles were where they are now.

Whatever were the mysterious mechanisms directing the swing of climate to and from glaciation, they presumably are operating still. The difference we experience in snow depth or temperature between one winter and the next is a difference of greater magnitude than that required to alter the glacial or non-glacial direction of our climate if sustained.

After considering the earth-rocking events that possibly were causes of glaciation in the deep past, we are still as puzzled as ever about why the glaciers came and went, but it will be exciting to watch how future discoveries shed new light on this baffling enigma.

CHAPTER 8

FROM MAN-APE TO MAN

Wits may allege that speech was invented so that man might talk about the weather. If there could be any justification for such a foolish suggestion, it is that when language was being invented the weather was certainly something to talk about.

WILLIAM HOWELLS

WHEN the Pleistocene drama was beginning there were, in Africa and at least the southern parts of Europe and Asia, various types of apelike creatures that were acquiring Manlike features. Some gradually evolved to naught but others seem to have been clumsily on the way to becoming human.

Mild conditions of the Tertiary carried over into Early Pleistocene, so even southern Europe was still subtropical though temperatures were gradually falling. Glaciers already existed in some cold mountains and polar places, the oceans were cooling, and habitats were slowly changing with respect to climate and vegetation. The small drop in temperature felt in the tropics apparently was not an upsetting factor in those amply warm regions, but increasing dryness was. Desiccation was severe in parts of Africa, the continent where many remains of the earliest-known Manlike creatures have been found. A critical factor in their development seems to have been the spread of deserts and the shrinkage of forests.

They are regarded as man-apes for they embodied characteristics of both these animals, and are known as *australopithecines* because their fossil bones were first unearthed in South Africa (*australo*— south; *pith*—ape). Later digs found them farther north too. There are hints of their presence in southern Eurasia, but it is in Africa—

mainly southern and eastern Africa—that a rich and varied assortment of their remains have been discovered by Dart, Broom, Robinson, the Leakey family, and others.

South Africa was drier when the man-apes lived there, with the Kalahari being a larger desert than it is now. In East Africa along the Great Rift Valley much faulting and volcanism was taking place. Earthquakes shook the land, fiery lava poured up through fissures, and often volcanoes erupted violently. As volcanic ash fell it sometimes buried bones and materials associated with them, and thus preserved them. This area is of special interest because some of the volcanic material that contains potassium can be dated by the potassium-argon method, and so dates for fossils can be arrived at. Lavas that recorded the reversal of Earth's magnetic field also help in the fixing of dates.

When we first meet the australopithecines there were no glaciers in the African mountains, but these important characters in our story will be around to see glaciers hanging part-way down the slopes of Mount Kenya, Mount Kilimanjaro, the Ruwenzori peaks (the "Mountains of the Moon"), and other mountains as the Ice Age progresses.

The man-apes must have been in Africa (and perhaps elsewhere) as far back as five million years ago. Maybe longer. Some of their remains appear to be about four million years old, and they seem to have started making crude tools about three million years ago. They evolved in our direction with a dragging slowness that makes us appreciate the value of our heritage of intelligence and culture, for if we should lose it how long would it take us to recover it!

The timing and events are blurry. And the reader is reminded that dates here and elsewhere in this book are only tentative estimates, the best we have now. In areas where no chronological reference points are available, dating is especially speculative.

On the basis of current fossil evidence, it seems that two general types of australopithecines wandered from the Tertiary into the Pleistocene. There was the larger, robust type, *Paranthropus*; and the smaller, gracile *Australopithecus*.

The *Paranthropus*-type man-apes were the more apelike. They looked much like gorillas but walked erect (though the earliest types were probably knuckle walkers). They seem to have remained tree climbers too, for their lower limbs were rather short and their big toes were somewhat like thumbs. Adult males were about four and a half to five feet tall and weighed between 150 and 200 pounds. They had long, heavy skulls and massive jaws with powerful molars, nearly twice the width of ours, needed to chew nuts, berries, leaves, bulbs, and roots. They lived mainly on vegetation, only rarely seeking small game. Their teeth were not adapted for killing or meat eating. Above their eyes the braincase was low, so they had no forehead. Their skulls have the sagittal crest on top, the primitive front-to-back ridge to which heavy jaw-working muscles were attached. The best defense of these robust man-apes must have been that of the gorilla—a fierce appearance that frightened off attackers. They apparently lived where climate was fairly moist, where there were trees and where vegetation was plentiful. They were around from Late Tertiary to mid-Pleistocene, by which time they were living alongside true men. And then they disappeared or are lost track of. Stone tools have been found with some of their remains, but since these herbivorous man-apes seem not to have cut up animals for food, it is assumed the tools were made by someone else. But maybe not.

Man-apes of the other type, collectively called *Australopithecus*, were smaller and of lighter build, and closer to the path that led to Man, if indeed they were not directly on it. The average adult female was only about three and a half or four feet tall and weighed forty to fifty pounds. She was slender but had wide hips. Males were a little larger and probably more heavily built. As time went on these man-apes gradually increased in size. They were less fearsome-looking by our standards, and in that respect were even more defenseless than their larger, more gorilla-like "cousins." They lived in a drier environment, in the grasslands and scrubby open woodlands, exposed to predators. They themselves were predators. Unlike *Paranthropus* but like ourselves, their teeth were suitable for eating meat as well as plants. Their legs and feet were similar to ours. They not only walked erect but could stride and run as we do. Their facial features were partly Manlike, partly apelike. The mouth area protruded, but not as much as an ape's. The nose area was flattish. They had the beginning of a forehead, but their brain was still in the ape-size range with an average volume of about 450 cubic centimeters (about the same as *Paranthropus'*). A chimpanzee's is about 400 cc., and a gorilla's about 500 cc. Modern Man's averages about 1,450 cc.

Many fossilized bones of *Australopithecus* were preserved in limestone caves of South Africa. They may have been washed into the caves during rains or dragged in by animals. There and also in sites farther north were bones and shells of contemporary animals which probably served as part of their diet: baboons, hares, bats, birds, moles, tortoises, crabs, lizards, fish, porcupines, hyraxes, and hogs. There were larger animals in their territory too—the wildebeest and other antelopes, the giraffe, rhinoceros, hippopotamus, elephant, horse, crocodile, hyena, saber-tooth tiger, lion, and leopard —but probably the man-apes usually caught only the young or disabled of larger animals. For good reason they did not go after the predators. Plant foods and eggs were also on their menu.

Sometimes a group of *Australopithecus*-type man-apes drove large animals into water or a swamp and then stoned or clubbed them. (Rock throwing and clubbing persist in our "civilized" society. We are reminded here what level of intelligence and aptitude they represent.)

Australopithecus may have used any available object for a tool or weapon. Sticks must have been used to dig up roots or reach rodents in their holes. Baboon skulls had been crushed by blows, probably made with the long bone of an animal, so the tasty morsels inside could be eaten. Tools made of crudely broken rocks and bones have been found along with *Australopithecus'* remains but we cannot be sure he used them. We suspect he did.

It would seem that *Australopithecus* used tools and brandished weapons, for he had no built-in tool or weapon such as other animals have. He almost had to have some artificial means of defense and offense to survive in open country, and some way of disassembling his animals. Killing an animal is one problem, but getting it ready to eat is a bigger one. Ask any hunter. How could *Australopithecus* slit the hide, loosen the flesh, hack bones apart so the meat could be carried away or passed around, just with fingers and teeth? There had to be a better way. He was hungry and in a hurry and so were his companions. He picked up something that could do the job, a hard rock with a sharp edge. Not any rock would do. Soft sandstone or crumbly shale, for instance, would be useless. But a hunk of obsidian, which is volcanic glass, or chert or flint or quartz or quartzite—these would do nicely. (Chert and flint are rocks of the same composition but chert is light-colored and flint is gray.)

So as a start to using tools, sharp rocks were picked up when needed and dropped as the user moved on. Eventually, perhaps after a half million years, some genius thought to break one rock with another and so make his own cutting edge when he could not find a handy rock already broken. Maybe a few hundred thousand years later some other genius thought to crack several chips off such a rock to make a more efficient tool. The idea caught on and spread, for aping is one of the primate traits. But improvements came extremely slowly.

It is difficult to tell whether a crudely broken rock with certain characteristics was broken by nature or by hands. Between naturally broken rocks and the first ones known definitely to have been cracked by hand are *eoliths*, or "dawn stones," which only vaguely look as though someone had struck and broken them. They do not show recognizable workmanship as those rocks prepared by more advanced toolmakers do. At what appears to have been a kill site or camp an archeologist searches for evidence that someone had been there, and he finds these eoliths. They are of the size and shape that could be held in the hand and their broken edges are sharp enough to have shaped a stick or butchered an animal. Still, the edges could have been chipped as the rocks fell off a ledge, or were banged about in a rapids, or tumbled in an avalanche, or were tramped on by animals, or cracked in frost where there was frost. He sees many ordinary chipped rocks around but something about a certain one intrigues him. Say it's a hard, hand-size rock worn smooth by long grinding in a stream bed or along a shore. A pebble he would call it, and one side is cracked off. He picks it up, and an eerie sensation darts through him, for the stone is exactly the right fit for the hand and when held comfortably the sharp edge is just where it should be to do its cutting or chopping.

This rock the archeologist holds could have been one that had never been touched, but there are marks where the rock was struck. Even so, those also could have been made naturally. He can't be sure. But there are times he sees an eolith and feels sure it was "made" by someone. It is a type of rock not found locally, so it must have been carried there. It is found with a scattering of bones, and the bones were smashed open for their marrow and show signs of being gnawed. Someone had been there. Sometimes there are many rocks of the same composition with flakes chipped off. When a number of rocks are found, all having been chipped in the same fashion, there is no doubt that Early People were making tools, and at that stage we definitely have Man. But who is he?

If *Australopithecus* made tools we can cross the "ape" out of "man-ape" as far as he is concerned. But maybe he was the prey of another type of man-ape or of a superior "somebody." Small of body and brain, helpless-appearing as he was, he survived in an environment filled with dangers from Tertiary time to the middle of the Ice Age. No small accomplishment. Then he left the stage for good. He or something like him had evolved into a higher primate, Man.

There is a gradual change of cast through Early Pleistocene; and in early Middle Pleistocene *Homo erectus*, erect Man, takes over the stage. He was so named when it was thought he was the earliest Man to walk erect. Later it was learned that other Early People before his time had acquired that posture. He appears in a number of forms, as did the man-apes. *Homo erectus* is an overall term that includes many similar or related types. There is much difference of opinion as to which specimens—some consist of just a piece of skull, or a jaw, or a few other bones and teeth—should be included in the *Homo*, or Man, genus and which should not.

Some anthropologists believe that a species of *Australopithecus* evolved into *Homo erectus* and perhaps through him to *Homo sapiens*, wise Man—us. But others believe that along with *Australopithecus* there lived still another unknown kind of near-Man who was superior to him, and it was he who made and used the rudimentary stone tools, and it was he who was responsible for *Australopithecus'* bones being found with these tools; and they think he may have been partly responsible for the disappearance of Mr. A and other man-apes, having looked upon them as enemies or food.

Just as we have seen the various man-apes supplanted by *Homo erectus*, so we shall see *Homo erectus* in his various forms being supplanted by Modern Man. Through what evolutionary channels we cannot say. By whatever cross-breedings, mutations, genetic drift, or assimilations that occurred along the way, some populations have been swallowed up by others, figuratively and to some extent literally perhaps, until eventually we are left with our lone remaining species of all the many that began the shuffling march toward Manhood.

Were we the most intelligent, or the most prolific? Were we the best killers, or were we the meek who inherited the earth? We have to believe we were the best tool and weapon makers as we survey the plethora of machinery and gadgetry we have produced. And our weaponry is so "superior" that it made us victorious and sole survivor of all rival species. Ironically, it could turn out to be what causes our extinction and thus the annihilation of everything human. That seems all too possible when we consider that the Ice Age was a time of mass extinctions of large animals and that many types of Early People vanished completely during that time. Now there is no other species for us to war with, so we foolishly tempt fate and court total death by warring with each other. There is no reason Modern Man should be the safe exception. The setting again could be the Tertiary. Only this time there would be no one to start the long climb over again by banging a rock to get a sharp edge. There are no man-apes left. No living apes or monkeys were ancestors of ours; and they have gone too far in a different direction to ever be able to retrace our steps.

The recovered remains of Early People are so fragmentary and diverse, and come from such unconnected places, and are so poorly dated in most cases that we can only guess what their ages are and who was a contemporary or relative of whom. As someone said, we are not looking for missing links. We have some of those. We are looking for the chain.

Homo erectus represents a major stage in the development of Man, having reigned apparently through the third-last glacial stage and part of the second-last interglacial. Remains of this kind of people have so far been uncovered in places around the Mediterranean, in East Africa, Europe, northern and western China, and Java, indicating they were widespread throughout the Old World. However, central and far-northern Eurasia yield no clues of Early People up through Homo erectus' time and are believed to have stayed largely unoccupied until the last glacial stage. No one had yet shown up in the New World or Australia. The territory known to have been occupied during Homo erectus' time consisted of Africa, except for the deserts and dense rainforests; central and western Europe, including southern Britain; and the area through southern Asia to northern China. The total number of people in this territory was small, probably only as many as are living today in one of our larger (but not largest) cities, and they were thinly distributed through the area.

Although the various skulls assigned to this species show dissimilarities, they are homogeneous enough to be grouped together until distinctions can be made. The variations in skull shapes and body forms in any gathering of people today—at the United Nations or in your home town—would probably show differences at least as great as those within the species *Homo erectus*. Those people had traveled widely and mixed well.

Whereas the man-apes may have roamed to the edge of the sub-tropics or a little beyond, their successors ranged farther north and became entrenched in middle-latitude regions with cold winters. People have been there ever since. The "erect people" experienced at least one glacial episode. In the mountains of Asia were many glaciers and ice caps which they probably encountered but avoided if they wished. However, in Europe it was less easy to escape from the ice and its effects. The ice sheet hung over the north and let its cold spill over the south, and mountain glaciers were close everywhere. There the third-last glaciation (the Mindel) was the greatest one of the Ice Age, with the ice sheet reaching its maximum extent. The ice front crossed the Netherlands, central Germany, and southern Poland; and it abutted against the northern Ural Mountains, which separate Europe from Asia. Ice covered Britain south to the Thames, and almost completely covered Ireland. When the ice sheet spread out of Scandinavia and glaciers crept down from Europe's many mountains, animals and whatever people were on the scene moved south to the Mediterranean and to North Africa, which was a refuge in glacial times. But during interglacials and warm phases of glacial stages the people and some animals typical of warm regions moved back north into Europe as climate improved.

North Africa in glacial times was sometimes rainier than now, a far more inviting region for nomads living off the land. During pluvial, or rainy, times the desert shrank, and grasslands or Mediterranean-type woodlands covered much of the land that is now arid waste, particularly on the periphery and in the highlands of the Sahara. So during certain periods of the Ice Age that desert was not the extensive barrier it is today. People and animals could live well in much of that region and easily move across North Africa. Fish, crocodiles, and hippos were present there in fresh-water lakes and streams that are now gone, and their bones are found along with those of *Homo erectus*.

When the ice was broad and thick upon the land, sea level fell, and this may have made migration to Africa easier than is now possible.

A land connection may have existed then between Europe and Africa but this is uncertain.

Lowered sea level exposed broad continental shelves off western Europe and southeastern Asia and elsewhere. It was probably during such a time, when islands off Asia became part of the mainland, that *Homo erectus* made it to Java. It is believed that he was already there in one of his more primitive forms, referred to as Java Man, during the third-last glacial stage.

Members of the species *Homo erectus* were about five feet tall, taller than *Australopithecus* but shorter than the average person today. Their brains were larger than the man-apes', ranging from about 775 to over 1,100 cubic centimeters, the largest size being within the lower limits of Modern Man's brain size.

Homo erectus not only stood and walked as erect as we do but he had a skeleton indistinguishable from our own. Only his head differed somewhat from ours. Skulls or parts of skulls from different times and places exhibit regional variations and development through time. The earlier representatives of the species, including Java Man, had skulls that were thick and heavy, and large, robust jaws. Their faces had strong features. Their teeth were much like ours but a little bigger, they had powerful neck muscles, and above the eyes were bony bulges, or brow ridges. The top of the head was still low and flattish with the sides of the skull tapering in just above the ears, for the brain was relatively small. No Early People had yet developed a real forehead or chin.

These erect people were proficient hunters now, even of large animals. They probably lived in bands, staying around a home base near water for some time and then moving on. Tools improved as the brain improved. In central Europe and southeastern Asia the principal tool was still the primitive pebble chopper, made by crudely breaking off or chipping one end of a hand-size rock to create sharp edges. But in western Europe and southwestern Asia and Africa a more advanced tool was being used, the hand-ax—a poor word choice, for it did not look at all like our ax. It was a more or less elongated piece of flint or other hard rock that usually had a rounded end (which the hand held) and a more pointed tip with sharp edges. Its typical shape was something like a pear or almond or teardrop. Hand-axes were chipped over most or all of their surface. Some were crudely, hastily made; others were handsomely fashioned.

They were multi-use tools, but almost certainly one of their chief functions was to pierce and cut animal skin and hide. A great many hand-axes have been found throughout the Old World, but the bones of those who used them are sparse. In Europe alone tens of thousands of hand-axes have turned up but only a few bones of their makers. Another popular tool was the cleaver, which had a straighter cutting edge and was more like our ax (but without a handle) than the misnamed hand-ax.

The largest assemblage of *Homo erectus* remains was dug up in a limestone cave twenty-seven miles southwest of Peking, China—parts of over forty men, women, and children. When and how long these Early People lived there at 40° N is unclear, but it seems they were there during the third-last glacial stage and second-last interglacial.

In caves of that area were bones of bears, hyenas, and several members of the cat family, all of which must have challenged the people for possession of those shelters. There were also bones of deer (apparently the main food animal), horses, camels, elephants, antelopes, buffaloes, rhinos, and sheep, probably the remains of meals eaten there.

Peking Man had made some anatomical advances over earlier members of the species. His brain was about 200 cc. larger than Java Man's and the top of his skull had risen. Besides the pebble-chopper tool Peking Man had stone scrapers and points and used animal bones as tools. He may have at times eaten human flesh. There was an abnormally large proportion of head parts among the cave bones, suggesting he brought heads there so he could leisurely nibble away at their insides. Some anthropologists do not accept the cannibal interpretation since Man does not kill his own kind for food, except in rare circumstances such as when he is starving or, among primitive people, in superstitious rites. (Primitive people have been known to eat the flesh of a person who has a trait they admire, believing they will thereby acquire that trait.) Or there is always the possibility that another kind of Man, superior or different, may have looked upon Peking Man not as a species-brother but as a lesser game animal.

One impressive, huge animal that lived in the Far East during the days of *Homo erectus* was *Gigantopithecus*. At first it was thought to have been a giant Early Man, but more study revealed it to have

been an advanced, over-size ape that may have weighed as much as
600 pounds. Archeologists had their first clue to this extinct ape a few
decades ago when they found its enormous fossil teeth being sold
in Chinese drugstores as dragons' teeth to be ground up for medicine.
Some molars were six times as broad as ours.

In East Africa during part of *Homo erectus'* existence lived some
other immense animals: a pig as large as a rhinoceros with tusks so
gigantic they were first thought to have belonged to an elephant;
sheep that were six feet high at the shoulder and had horns with a
span of twelve feet; baboons larger than gorillas; and ostrich-type
birds that stood twelve feet tall!

We still have not mentioned the most outstanding advance displayed by *Homo erectus*—the use of fire. Whether he was the original
master of fire or acquired it from someone else is not known, but we
see its controlled use first, definitely, with him—the ashes, charcoal,
and burnings at his hearths and campfires. The taming of fire, Man's
first energy other than his own muscle, was as important a step forward as the first breaking of a rock to make a tool. He could not have
moved into winter climates without it.

Traces of fire were found with Peking Man in northern China, and
at a number of sites in Europe. In Europe it helped people endure
the third-last and most severe glaciation (maybe about a half to three
quarters of a million years ago). Hearths were found on the French
Mediterranean coast, and charcoal with charred bones in Spain;
and near Budapest, Hungary, an old hearth site exposed in a quarry
wall contained the split, burnt bones of deer, bear, wild ox, horse, and
dog, as well as some pebble tools and a few teeth, unmistakably those
of *Homo erectus.*

Before people learned how to control and make fire it was something to fear and flee from or to be fascinated by from a safe distance.
How many times it must have been daringly captured and held like
a torch or carried away to be studied or played with dangerously.
It could cause pain or rush out of control. A breeze could blow it out
or whip it into a holocaust in dry country. With increased intelligence
and experience people learned how it could be fed and confined and
made useful. Before they accidentally learned how to create fire by
striking pyrite with flint or by continuously rubbing sticks together,
it was obtained by snatching burning or smoldering wood or other
objects from fires that had been caused naturally by lightning strik-

ing dry vegetation or oil seepages, by fiery lavas, or by spontaneous combustion in coal beds or oil shales.

Fire radically changed the lives of those who possessed it. It was comfort, allowing them to live year-round in regions with cold winters even before they wore clothes. It was a defense, for animals fear it. For the first time a person, even a child, could stand up against a predator and hold his ground. With fire one could chase animals out of caves and use them for his own shelter. Fire was an offensive weapon too. Bands of men armed with torches could drive animals into a swamp or corral, or stampede herds over a cliff, so meat in large quantities became easier to obtain. This was important in winters when plant food was scarce. Fire was illumination, and where winter nights are long, far from the equator, people could continue working after twilight. Though only the lighted, airy openings of caves were lived in, now with fire the deep recesses could be explored. Fire stirred the imagination, and as these people stared into its flares and dancing shadows feelings of the supernatural must have been ignited and intensified. Fire on the hearth created the family circle and increased social cohesion. Fire helped in woodworking. Fire cooked food, so now the meat and plant foods were more tender, digestible, and palatable. Less chewing was required and with time the size of the teeth and jaw would diminish, and the muscles that worked the jaw would become smaller. The skull would become lighter and thinner. It was easier for the brain to enlarge, and it did.

The erect people had spread north, and with their improved tools and fire many stayed there during the coldest of all glaciations. Though the nights and winters were cold and the chill of the ice was around them, they could be comfortable and healthy and could eat well. However, without clothes they could not yet live in the tundra bordering the ice. Of course, during parts of the interglacials, northern regions were warmer than they are now.

Did Early People who lived in cold climates have more hair on their body than we do? There is no way to tell, of course, unless a preserved body with hair still on it turns up in the Siberian permafrost. Since we have light hair and hair follicles over our bodies now—and some people are quite hairy—it would seem that people could have had more hair in the past. As naked people moved into cold climates perhaps they developed an insulating cover of hair just as the woolly mammoth and woolly rhinoceros did. If all we had to go by were the bones of those animals and their nearest living counterparts, the modern elephants and rhinos, we would think it unlikely that they ever had much hair; but frozen specimens of these animals were discovered with thick hair still on them. Many animals grow heavier hair or fur for warmth in winter. On the other hand, maybe we began our existence with a hairy body and lost the hair later. Or, another possibility: maybe our lack of body hair gave us the incentive needed to develop our latent abilities by taming fire, making clothes, and building shelters.

Culture became increasingly more important to Early People as a controlling factor in their survival. They could no longer live without it—without their tools and weapons, without fire, without group cooperation. Family life had begun by now if not earlier. Its protection was necessary, for as the brain and head were enlarging, a child had to be born at a more premature stage. The baby that stayed in the womb too long never made it, and neither did the mother. Helpless infants had to be cared for until they were strong enough to fend for themselves, so the mother (pregnant much of the time) was nurse and child watcher while the father was defender and hunter. People must have had language, though simple, by the time they lived cooperatively in groups, hunted together to kill dangerous animals, and had brains three fourths the size of ours.

As time passed, people's physical endowments would become relatively less important to their well-being and survival, and they would grow increasingly reliant upon artificial devices and the group's accumulated knowledge.

Theodosius Dobzhansky, zoologist-geneticist, has written of today's world: "There is no doubt that human survival will continue to depend more and more on human intellect and technology. It is idle to argue whether this is good or bad. The point of no return was passed long ago, before anyone knew it was happening." That point of no return may well have been in the days of Homo erectus—if not before.

There are so few remains of Homo erectus in spite of his long tenure on Earth that we see only fleeting glimpses of him in his various likenesses as he changed over many millennia and migrated across the continents. During the next long, empty period information about him and his successors is scanty. There came another slow surge of ice (the Riss), and by the time it left, more advanced people had arrived—among them, Neanderthal Man.

CHAPTER 9

THE STRANGE FATE OF THE NEANDERTHALS

By speeding up the centuries I could visualize a tidal wave of ice flooding down from the Arctic and crushing everything before it. . . . I could see nothing but the obliterating ice, hear nothing but the wind, and feel nothing but the rigidity of death. But, finally, I could see the ice imperceptibly sinking; and the ocean rising as the ice melted; and the land resurrecting under the sun, with the mountains scoured and planed, and the rivers pushed into new courses. And along the edges of the land in Europe and Asia I could see men with primitive tools laying the foundations of history.

RICHARD E. BYRD

THE TIME was Late Pleistocene. It was the last inter-glacial period and the spotlight was on Europe. The Early Neander-thals entered the scene inconspicuously, perhaps 100,000 years ago, and until about 35,000 years ago they would be developing into a most distinctive type of human being.

There were Neanderthal types living outside Europe too—in North and East Africa, the Middle East, southwestern USSR, and eastern Asia—but not so much is known about them; and their Nean-derthaloid features were usually diluted with those of other people. It is in Europe that the Neanderthals attained the classic, extreme form for which they are famous. Other types of people seem to have been present there also, but they have not shown themselves clearly as the Neanderthals have.

It was somewhat warmer than the present in that last interglacial before the last glaciation, the Würm, which was due to begin about 75,000 years ago.* Southern Europe again had subtropical Mediter-

---

* "Würm" here will apply to the last glacial stage throughout Europe, although that term, strictly speaking, applies only to the Alps, because to use other regional names would confuse the lay reader and because the general literature and that dealing with Early People usually uses "Würm" in a broad all-Europe, even all-Old World, sense.

ranean climate and vegetation, and most of the rest of continental
Europe including southern Britain was reclothed in temperate forest,
mainly broadleaf with some coniferous trees, with meadows inter-
spersed. The ice was gone, more "gone" than it is today, so the seas
had risen even higher than they now are. Because of that and because
the Scandinavian Peninsula may have been still depressed from the
weight of the former ice sheet, the sea for a time not only covered
the present straits and bays between that peninsula and the main
continent but also made a strait between the Baltic and White seas,
leaving Norway, Sweden, and most of Finland as an island. That
Fennoscandian island was covered mainly with boreal coniferous
forest as was northern European Russia. The British Isles were separated from the mainland as they are now. The Hungarian Basin and
the region north of the Black Sea were dry steppe.

In the forests were the straight-tusked woodland elephant, the
woodland rhino, a stocky horse, the ox, moose, wild boar, and several
kinds of deer. On open, grassy plains were giant deer and horses.
Hippos wallowed in streams and ponds. Caves were the dens of cave
lions, cave leopards, hyenas, and bears. In the mountains were the
ibex, chamois, alpine wolf, and marmot. Also among the animal
population were bison, beaver, lynx, weasel, and marten. Fish and
shellfish were available. Far to the north in the tundra, marshes, and
swampy forests of elms, willows, and pines were reindeer, muskoxen, and mammoths.

The deep heartland of Eurasia and the Arctic North were still
unoccupied by Man, as far as we know. Evolutionary changes were
occurring in the Far East, for we see the results later in the populations that developed there and in the migration to America which
would begin during the next glaciation, if not before; but archeological evidence is scarce thus far and we are still much in the dark as to
what was going on there.

Neanderthals on the western European peninsula seem to have
developed their aberrant, classic physical characteristics due to con-
finement. They were so unusual that when a composite picture of
them was first pieced together from their bones found throughout
Europe they were looked upon as a kind of stooped and stupid freak.
It took a good while before their status was cleared and they were
given the respectable rank among Early People that they now hold.

Water restricted their movement on the south, west, and north.

The high interglacial sea level helped isolate them. Any land bridges to North Africa that may have existed would have been submerged again, though, of course, the circuitous route via Asia Minor and the eastern rim of the Mediterranean always stayed open. Conditions in North Africa had deteriorated again as drought returned to some areas, lakes and streams dried, and much good grazing land reverted to desert. The Sahara again became difficult to cross.

No Neanderthal remains have been found in Britain, but artifacts of their time are there.

The mountain ranges that subdivide Europe did not constitute real barriers to travel but did impede it. Later when glaciers grew out from them many passes and corridors were blocked. The eastern steppes that extend into Russia were not especially inviting, having a smaller variety of game, less water, less protection, strong winds and blowing dust, and little wood for fires and implements. So the European Neanderthals were hemmed in, and the relatively rapid climate changes before and during the Würm glaciation, the numerous temperature and moisture fluctuations, and the attendant environmental upsets had their effect upon processes of natural selection and helped cause this localized, inbreeding population to develop a unique physique.

Gradually the climate cooled. Mountains became capped with snow and stayed white all year. Trees in the highlands died and those on the plains were thinning and changing from warm-climate to boreal types. The snowline crept lower and tundra vegetation which had taken refuge in high elevations during the warm interglacial moved downslope ahead of the line of perpetual snow. Glaciers reformed in hollows among the mountains and with stealthy, unnoticeable slowness inched down the slopes. They grew in the Alps, the Pyrenees, the Carpathians, the Apennines, and in other highlands of France, Spain, Portugal, Britain, central Europe, and the Balkans.

The climate went through many oscillations of warming and cooling but the chilling trend was persistent and intensifying. More lakes and rivers froze in the winter. More summer days and nights were cold. Fires burned brighter and longer on the plains and shores and at rock shelters. Some edible plants were not growing and producing as well. Some animals migrated farther south each winter and did not return as far north in summer. During the beginnings of a glacial stage Early People would normally be expected to drift to a warmer region, but this time a new breed was breasting the cold. The Neanderthals did not flee.

Notwithstanding all the jokes that have been made about Neanderthals, and the unflattering pictures that have been drawn of them by artists and writers, they must be admired. They were not beautiful people, but beauty is not what counts when the tally of achievements is added up. They were rugged, adaptable, unyielding. They hung on as the cold worsened and the ice came, and would not leave. Their bodies changed, and their way of life, as they shared Europe with the spreading ice but held their ground tenaciously. Could, or would, modern people do the same? Some Early Neanderthals undoubtedly did leave when the climate turned cold, but they did not make history. Presumably some non-Neanderthals stayed on along with the Neanderthals, but they have hardly been heard from and left only faint marks on the record. More news of them may come to light later. In them (perhaps in Neanderthals too) may be a vital part of our ancestry.

Neanderthals had not been heard of either till the 1800's. The first fragmentary Neanderthal skeleton was discovered in 1856 in a valley grotto east of Düsseldorf, Germany. It set serious minds to thinking. The concept of evolution was not generally known then. Darwin's *On the Origin of Species* would not be published until 1859. And these bones turned up. Most peculiar. The person had been somewhat bowlegged and the skull showed heavy brow ridges. One examiner concluded he had been a horseback-riding Mongolian Cossack of the Russian army that drove Napoleon out of Russia in 1814. Others said he was the victim of rickets or some other bone-crippling disease, and that pain caused his brow to be misshapen. Others said he was an antediluvian specimen washed there by the Flood.

Interestingly, the valley in which he was found was named for one Joachim Neumann, who happened to like his name (which means "new man") in its Greek translation, Neander. *Thal* is the German word for "valley" in its old spelling, but it is pronounced "tahl," as in its modern spelling, *tal*.

More Neanderthal bones and skulls kept turning up around Europe, mostly in caves and under rock ledges. The interpretations given of them caused fireworks, for evolution was a controversial subject. As time went on a general picture emerged of the Neanderthals' build, facial features, and way of life.

Classic Neanderthal Man had a powerful, bear-like, barrel-chested build. He had heavy leg bones, somewhat bowed; large joints; and large feet and hands, but short fingers and toes. His forearm was a little short compared to ours, but made a good lever that way. The bones of his legs and arms show he must have had heavy muscles. He was a little over five feet tall, and Neanderthal Woman was a little under five feet.

The Neanderthal skull was large and so was the brain, but the skull had enlarged not upward as ours has, but backward, and it bulged out slightly on the sides. The back of the skull was elongated as though it had a pug at the back. The face was large and projected forward. The nose was wide, and above the eyes were heavy, bulging brows. The teeth and jaw were large; the hardly existing chin was receding. So Neanderthal Man had many primitive features and therefore was described for about a century as a club-swinging caveman, a dead end of an ape line. But as more evidence came to light and as scientific means of reconstructing past people and events improved, a new image of these people emerged. It is believed they had a native intelligence as great or, in some cases, greater than ours. They were sensitive people capable of abstract thinking and emotion, and skillful operators for their day.

Their tools and weapons had shown marked advances in a relatively short time as compared to the ponderously slow improvements made by previous people. Their industry is classed generally as the Mousterian culture. It may have belonged to other contemporary people as well but it is clearly associated with Neanderthals.

The pointed hand-ax was still in vogue in Early Neanderthal time but gradually it and other of the old heavy tools were replaced by more specialized and efficient lighter-weight ones. Easily worked flint was the most desired tool-making rock. When freshly broken its edge is as sharp as glass, and it is harder than any metal used before steel was invented. New techniques of striking flakes off flint nodules and edge-chipping were used to make knives, scrapers, and small points. The first known projectiles—hand-thrown spears—date from this period. Large numbers of tools were found near springs and along lakes, rivers, and coasts, these being locations favored by their users.

Hunting methods and weapons were improving, and hunting was

excellent. Bands of men working together outmaneuvered and killed the largest of animals. Pits were dug and covered over so animals would fall into them; and birds and small animals were trapped. Hunters very likely used the bolas, a device made of rocks tied to the ends of long leather thongs, which was swung at an animal with great speed, often at its legs to entangle them. The Neanderthals were highly proficient, well-organized hunters. Had they not been they could not have stayed on as the glacial climate returned.

The weather grew colder and damper. Round-topped deciduous trees died and pointy-topped conifers took their place. The landscape of much of western Europe was like the taiga forest of Siberia or Alaska. Then over the years those cold-resistant trees thinned and died, and earth-hugging tundra vegetation covered the land north of the icy Alps and also the highlands to the south. Temperate broadleaf and coniferous forests survived along the north shore of the Mediterranean; and during this first glacial thrust of the Würm, the hippo and rhino retreated to the peninsula of Italy, as far south as they could go. Cold steppe extended from east of the Alps across central Russia. Permafrost hardened the ground; and in summer only the surface thawed, allowing low vegetation to grow and forming marshes in poorly drained spots. Winters were bitter cold with winds strong from the regions of the ice.

With flaming torches the Neanderthals drove bears and lions from caves and took them for their own shelters. They kept fires burning at the cave mouths and they were safe. When snowstorms raged they stayed warm there behind windbreaks, huddled among animal furs. They watched branches of green spruce trees droop under heavy robes of snow. Ice glistened on the river, and rabbits and foxes made tracks over the white ground. At night as roasting meat sizzled on the hearth, lions snarled in the outer darkness and their eyes flashed with reflected firelight, but they dared not come near. And the Neanderthals squatted at the smoky cave entrance, eating heartily, flaking their flints, and planning the next hunt.

Many animals they had known became extinct. Some migrated south in winter and back north in summer, so there was a seasonal change of fare. Other animals adjusted to the cold, as the Neanderthals had, and stayed around to meet tundra animals that drifted south ahead of the expanding Scandinavian ice. Among these tundra animals were grazing herds of woolly mammoths. They were the size of the modern elephant and had a peaked head and humped shoulders. With immense curved ivory tusks up to thirteen feet long they brushed away snow as they ate the tundra grasses and plants. A thick undercoat of wool and an overcoat of long hair gave them a shaggy appearance. Their color was probably a rusty brown, though the shade may have varied over the body. Even their trunks were hairy. Their ears were relatively small.

The woolly mammoth can be described quite accurately because many carcasses of the animal in various stages of preservation have been found frozen in the permafrost of Siberia and Alaska, where they were accidentally trapped and buried. In Alaska, it is said, mammoth hair was so common in places that it was an annoyance to gold diggers. One Siberian mammoth that lived about 39,000 years ago still had the remains of its last meal in its stomach, which showed it lived mainly on tundra grass. Under the skin of these animals was a layer of fat several inches thick.

The woolly rhinoceros entered the scene too. It also has been found frozen in Siberia. It was much like the modern rhino but had dense underfur and long shaggy outer hair as the mammoth did. Its head was carried low and on it were two horns, one on its forehead and the other on its nose, the latter being about a yard long.

Reindeer were there too, and Neanderthals used their skins along with hides and furs to cover their bodies. Scraping tools tell us Neanderthals prepared skins to wear, and at least one skull had teeth worn far down, the way Eskimos' teeth become from chewing animal skins to soften them. But there is no sign of sewing implements or fasteners that would suggest they made actual garments.

European Neanderthals are the first people known to have lived under subarctic and arctic conditions. Those of the forest-tundra fringe apparently roamed in the tundra in summer, camping in the open or in temporary shelters as they followed herds of horses, bison, mammoth, and reindeer that grazed there. In winter they and the animals traveled south to Europe's partially forested, more protected, somewhat warmer regions.

Since many Neanderthal people lived in caves and rock shelters during cold weather, much of their life is recorded there in the layered accumulations of refuse and sediments. They even left a few footprints. These people were the first known to have buried their dead, and many of the graves were in the floors of caves. They seem to have believed in an afterlife, for sometimes stone tools and other goods were placed with the deceased. Some of the things they did showed tenderness, appreciation of beauty, and religious feeling. In their caves, in what may have been ceremonial or religious meeting places, they sometimes set the bones and heads or skulls of bears in certain arrangements or laid them in niches or stone chests. Cave bears may have been their arch enemy. They were among the largest bears that ever lived, rivaling in size the Kodiak bear of Alaska, which is the largest living terrestrial carnivore. The cave bear of Neanderthal times was mainly herbivorous, but still a dangerous beast.

In central Asia a Neanderthal-type boy was buried in a grave partly lined with ibex horns. In a cave in Iraq is a Neanderthal grave where 60,000 years ago a loved one was carefully laid to rest. Fragments of a pine-like shrub, bits of flowers, and pollen tell a poignant story, according to those who uncovered the grave. This body was lovingly laid on a bed of pine boughs and decorated with a multitude of flowers. The pollen shows the flowers to have been small and brightly colored and of many varieties. They apparently had been collected and brought there expressly for the burial.

Here also was a cripple with amputated arm who had been cared for from infancy until his death at about the age of forty, which is comparable to the age of eighty now. In Europe there were skeletal remains of arthritic and seriously wounded people who were also cared for in their old age.

A 40,000-year-old burial site in France, containing skeletons of two Neanderthal adults (presumably parents) and four children, shows the feeling of family togetherness and affection that existed then. The parents were buried head to head, and near the mother's feet in neat graves were the bodies of two children about five years old. Nearby was a slightly older child buried with two flint scrapers and a point, and covered with a triangular rock slab. A baby so tiny it may have been stillborn was buried there too, with three beautiful flints. All the bodies were aligned in the same direction.

No ornaments or drawings have come to us from Neanderthals. They probably did not have the leisure to make them. However, red ocher has been found at many sites, having been used both in powdered form and as a crayon. Since no drawings were made on the walls of their caves or shelters, it may be that they decorated their bodies with it.

The condition of the bones and teeth of Neanderthals showed they died relatively young. The oldest were about forty; a few may have been between forty and fifty. Most of those who lived past thirty were men. Life expectancy was shorter for women, very likely due to the hazards of childbearing.

Measurement of the large Neanderthal skulls shows that these people had a cranial capacity greater on the average than ours—1,600 cubic centimeters for theirs compared to 1,450 cc. for ours.

Put all the facts together and we see that Neanderthals were not the slouchy, ignorant brutes they were first thought to be. They were fully erect people with physiques stronger than ours. Their brain was larger than ours. They were expert hunters of the largest animals around and made marked improvements in workmanship on implements and weapons. They lived in the neighborhood of vigorous glaciers and lived well for their time. They were the first people known to have buried their dead and to show religious feeling and belief in an afterlife. They cared for those unable to do so themselves, and they appreciated beauty.

As the cold came their caves and rock shelters were protected by windbreaks of boughs or hides. On the plains where there were no caves or rock shelters completely artificial construction was necessary. While the ground was frozen sod-covered, dug-out shelters were impossible to make. In the Ukraine of southern European Russia has been found an old Neanderthal settlement apparently dating from the Early Würm glaciation. Still lying where they had been placed or had fallen were mammoth bones enclosing an oval area about nine by six yards on which there were fifteen hearths, thousands of pieces of worked flint, hundreds of bone fragments of various game animals, and a spot of red, ocherous pigment. This habitation was probably covered with poles and skins or brush held in place by the mammoth bones. If so it is the oldest known dwelling.

The northernmost evidence of human occupation during the Ice Age is along the Pechora River of northern Russia only 109 miles from the Arctic Circle, with artifacts of Neanderthal style. It may date from the last interglacial or later.

Considering all the facts, anthropologists finally paid Neanderthal people the greatest of tributes, assigning them to our own species, *Homo sapiens*.

After the Early Würm cold period, which began about 75,000 years ago, Europe experienced a warmer interlude from about 42,000 to 30,000 years ago. The interlude was not as warm as an interglacial but it was a comfortable respite from frigid glacial conditions and trees began to grow northward. During that time climate fluctuated, becoming warmer and colder many times, preparing to swing into the deep freeze of the latter part of the Würm. The Neanderthals had successfully weathered a glacial attack, so life for them should have been easier during the warm interlude. But here we bump into another of the Ice Age's exasperating mysteries. During the mid-Würm warm period, about 35,000 years ago, the Neanderthal people seem to have suddenly disappeared without a trace. As one anthropologist has said, it's as though they were "herded together and pushed over a cliff." After that in Europe—in fact, throughout the world—all known people were modern, like us.

The switch from a Neanderthal to a modern population in Europe actually must have taken centuries, but that is like overnight on the Stone Age calendar. In European caves where there was repeated human occupation there is a sterile layer of natural deposits between the last Neanderthal floor and the floor above on which the first newcomers lived, so there is no way of telling from that evidence whether the two populations ever met or not.

These modern newcomers, known as the Upper Paleolithic people (Upper—later; Paleo—old; lith—stone), were strikingly different from the Neanderthals. Their legs and arms were straight and proportioned like ours. They had brains almost as large as the Neanderthals' but the skull had expanded upward instead of backward, and it was thin and globular like ours. The upward expansion had given them a forehead. The facial features were delicate compared to the Neanderthals' coarse ones. The heavy brow ridges were gone; the mouth and jaw had sunken in, leaving a definite chin and pronounced nose; the head was held high. In basic structure they looked like Europeans of today. Men in some populations were over six feet tall and powerfully built, and might have resembled tall Scandinavians of today (provided the latter live outdoors and by the strength of their muscle).

What happened to the Neanderthals? Why did they disappear so fast, so completely? Did they become extinct before the Upper Paleolithic people came? If so, why, when the climate was warming? Or did the invaders cause the Neanderthals' extinction? Or is it possible that Upper Paleolithic people were on the European scene with classic Neanderthals all along but in such small numbers that they did not show themselves earlier? There are hints that this might have been the case. Could they have had a population explosion and assimilated the Neanderthals?

Some prehistorians believe there was no "sudden" break between Neanderthal and Upper Paleolithic people and culture, that a gradual transition took place. Others point out that the newcomers were not even partly Neanderthal. They were totally different. It seems there must have been some in-migration of Upper Paleolithic people in mid-Würm for the population change to occur in such a short time. If so, the invasion, even though gradual, could hardly have been accomplished entirely peaceably, because of the presumed difference in language and obvious difference in culture between the populations. The new people would have been competing for the best hunting and fishing sites, the best water sources and dwelling places, and with their superior weapons they would have won.

But where did these new Upper Paleolithic people come from; where did they evolve into their modern form and stature? This is another mystery that defies solution. If they were not in Europe all along, they probably came from the east. Some possibly from Africa. In a number of sites outside Europe there are remains of people who show signs of development in their direction, and there are signs of Neanderthalism as well. The answer is far out of sight.

Maybe the classic European Neanderthals did succumb to the fickle climate, degenerate, and die out. Maybe the invading modern people ruthlessly annihilated them. Or maybe they assimilated them. Let's be genteel and say Neanderthal women may have admired the tall, lanky strangers who gave them beautiful ivory and shell necklaces and utensils they never had; and maybe the Upper Paleolithic women admired the broad-chested, muscular physiques of Neanderthal men.

Anyhow, every now and then I see someone who, I swear, has Neanderthal characteristics. You probably do too. And I can't help thinking, maybe the Neanderthals live on.

CHAPTER 10

THE LAST GLACIATION AND ITS PEOPLE

Man need not despise his lowly origin. On the contrary, he has every reason to be proud of his descent from ancestors who never accepted life as a gift, but who were obliged, from the outset, to win their existence by their own perseverance, ingenuity and labour.

JOSEF AUGUSTA

WHEN the Upper Paleolithic people succeeded the classic Neanderthals in Europe, during a warm interlude of the Würm Stage, the climate was not much different from that of the present. But the worst of the Würm was yet to come. Then the returning ice sheet would be larger, the cold would be colder, than during the Early Würm, through which the Neanderthals had lived.

The relatively warm Europe that Upper Paleolithic people took over around 35,000 years ago must have been quite attractive, especially to any of them who may have come from the dry grasslands to the east. They may also have come from the Middle East or Africa. It is thought many did come from those dry regions because they were adept at carving bone, deer antler, and ivory from elephant and mammoth tusks. Apparently they had lived where there was a scarcity of wood, which is worked much more easily. They continued to use flint and chert and other hard rocks, as metallurgy was not yet known.

Europe was a green, moist area then with many lakes and streams, marshes, forests, and meadows. Here was the good life—an abundance and variety of food to be had by gathering and hunting. With much-improved weapons and tools, with fire, and with a relatively large population for that time, the day-to-day battle for survival was not as time-consuming as it had been for earlier and more poorly equipped people. There was leisure time now to express artistic inclinations. The result is seen in the fine workmanship of the Upper Paleolithic toolmakers, in their carvings, sculpturing of figurines, and their drawings done on walls of rock shelters and caves. Upper Paleolithic people were the first known artists.

Although theirs was a good life, it was still a hard one compared to ours. A study of some 76 Upper Paleolithic skeletons from Eurasia showed that fewer than half had reached the age of twenty-one, only 12 per cent were over forty, and only a few women had reached the age of thirty.

Technological innovation and advancement had begun with infantile slowness during the *Australopithecus* level, and increased in momentum as time went on—through *Homo erectus* and Neanderthal times and thereafter. Inventions, improvements in living, and ways of modifying the environment were being discovered at an increasingly faster rate. This acceleration continues into the present as our technological growth and environment modification race ahead with speed almost beyond our control.

As yet, Upper Paleolithic people did not have any more impact upon the environment than animals did, but they were approaching the cultural stage where they could be masters over the animals and where they would be making unnatural changes in the landscape with their tools and intentional burning of vegetation.

Upper Paleolithic craftsmen specialized in making long, parallel-sided, stone "blade" tools. They fashioned knives and scrapers; and chisel-shaped, gouging instruments for engraving or working bone, antler, ivory, and wood; and drills to bore holes in bone, shell, wood, and so on. Some implements were held in the hand and others were attached to wooden or bone handles. Articles for personal adornment could now be made.

Hunting was done on a larger-than-ever scale. Earlier people had learned to steer or stampede animals over cliffs and into corners or traps or other situations where they could be dispatched by a group of hunters, but Upper Paleolithic hunters became even greater mass killers of animals. Found at the foot of some Rhone Valley cliffs were the bones of more than 100,000 horses that had been driven to their death over a period of many years. Left at one encampment in Czechoslovakia were the bones of a thousand mammoths.

These were the first people known to have built "houses." About 30,000 years ago their semi-subterranean houses dotted the steppe of southern Siberia roughly along the 50th parallel of latitude from eastern Europe to the Pacific. Walls of brush and skins were held up with mammoth bones, and interlocking reindeer antlers were the roof framework.

Where caves were available they were used as shelters, mostly in winter. Otherwise outdoor camps were set up. But as the cold of the Late Würm intensified, caves were occupied more than before. They were plentiful in limestone regions of western Europe.

We know that Upper Paleolithic people wore clothing—not just skins and furs draped over the body or held with cords—for they had clasps, awls to poke holes in animal skins, and bone needles with eyes.

Cave housekeeping had shown progressive improvement during Neanderthal occupation, as seen in the remains left on cave floors at various time levels. New floors accumulated over older ones. And housekeeping improved even more when Upper Paleolithic people arrived. In some cave entrances, which is where the living quarters were, stones were laid like flagstone pavement. Sometimes several inches of ocher—often bright red—was spread over the floor like a carpet. It might have signified something more than cleanliness or decoration. Perhaps wealth or the performance of a ritual. Juniper torches gave light, as did lamps which were hollowed stone containers holding animal fat and a wick.

During the second half of the Würm the ice made its last great thrust in Europe, beginning about 25,000 years ago and reaching its maximum extent about 20,000 to 17,000 years ago. (The reader is reminded that all prehistoric dates are tentative.) It covered more of Europe than at any time since the third-last glaciation (the Mindel). Winters were Arctic cold; summers were fairly comfortable over much of Europe. The climate was generally drier than the Early Würm of Neanderthal times.

Again, sea level dropped as the ice grew, causing the North Sea and English Channel to become dry land, as did the continental shelf off France and around Ireland. The westward retreat of the coastline lessened the moderating marine influence over Europe as a whole and permitted greater temperature extremes.

The Scandinavian ice sheet had had a good start during Early

Würm, spreading outward from its source, the mountains along the boundary of Norway and Sweden. How far it spread then is not clear because the larger Late Würm ice sheet overran and erased most traces left by it. During the mid-Würm warm interlude the ice sheet shrank considerably but probably did not disappear altogether. It fluctuated in size during many climatic oscillations but finally in Late Würm it made its big move, covering all of northern Europe.

Powerful valley glaciers spilled rapidly down the steep western slopes of the Scandinavian Peninsula, deepening fiords along Norway's coast. But the Atlantic Ocean limited the ice's expansion in that direction. However, across the North Sea floor, from which the sea withdrew, the ice moved without obstacle until it met the smaller ice cap coming toward it from Britain.

The southern margin of the Scandinavian ice sheet ran south through Denmark (which its moraines helped make) and followed the Elbe River valley to south of Berlin; then it curved northeast through Russia to the White Sea. Toward the continent's interior the ice sheet received less moisture, so it was thinner as it spread into Russia. How far the ice sheet reached across the Arctic Ocean floor is not known, but it may have linked up with the ice caps on large islands to the north.

The ice sheet grew thickest over the Gulf of Bothnia, which is east of Sweden and not far from the original source region. There it was about 9,000 feet thick. The sheet's average thickness was probably about 6,250 feet.

The smaller, shallower British ice cap formed from the uniting of ice from the Scottish, Welsh, and Irish highlands, and its flow was controlled by the shape of the land underlying it. During Late Würm the outermost edge of the British ice cap extended from Limerick in western Ireland to southern Wales, curved north to near Manchester, and entered the North Sea at Hull in east-central England.

The ice cap over the Alps was made of valley glaciers that had enlarged and merged. Piedmont glaciers bulged out at the foot of the mountains. South of the Alps, glaciers advanced to the Po Valley, indicating that a main source of moisture was the warm Mediterranean.

Other places in Europe with large glaciers were the Pyrenees, the northern Urals, the Caucasus of southern Russia, Iceland, the Faeroe Islands between Iceland and Scotland, Novaya Zemlya, and adjacent parts of northern Russia. Small glaciers existed in highlands of France, Germany, Spain, Portugal, Italy, central Europe, and the Balkans.

Much of Europe's drainage was upset by the ice. As the ice scraped over the land and deposited drift, it obliterated old river courses. Its meltwater created new channels. There was much flooding and many lakes, and outwash covered wide areas. When bare outwash flats, muddy floodplains, and fresh till plains were dry, wind swept up great clouds of silt, carried them away, mainly eastward, and deposited them as *loess*. Today, this fine, wind-blown dust still thickly covers much of central and eastern Europe. Loessial soils—highly fertile, arable, and well drained—were among the first to be cultivated when agriculture began there, and continue to be excellent farming land.

At the height of the Late Würm glaciation, tundra conditions prevailed in much of unglaciated western and central Europe and northern Russia. This was a permafrost belt, and there were no trees. But over the tundra grazed large herds of woolly mammoth, woolly rhinoceros, reindeer, and other animals, so the hunting people stayed there. South of the tundra there was grassland, combined forest-tundra associations, or mosaics of steppe vegetation and mixed evergreen and deciduous forests. Near the Mediterranean permafrost was generally absent, and there temperate forests predominated. Hungary, Romania, and southern Russia had only a month or so of frost-free weather during the year, and winters were about as severe as the tundra near the ice sheet.

In North Africa the Rif and Atlas mountain ranges held small glaciers, as did some mountains of the Middle East. The Sahara region was relatively moist during the early part of the Würm, but cooler and drier climate prevailed during the latter part.

With ice proliferating and temperatures falling in and around Europe, one would have expected the cultural level of Upper Paleolithic people to deteriorate. But that was not the case. It was reaching a peak that had not been attained anywhere, any time previously —there right next to, amid the ice. Just how high a cultural level these people had reached was not realized until recent years.

Their carvings and sculpturings were found first, but their finest surviving accomplishments were secluded in caves and waited much longer to be discovered.

During Late Würm they were drawing on bare rock surfaces of caves and rock shelters, the only "canvases" available. This wall art reached its zenith about 15,000 to 10,000 years ago. The subjects depicted were mainly animals. Humans were drawn only infrequently and were merely caricatures. But the animals were amazingly true-to-life. Animals were, after all, these people's most valuable resource. Their existence depended upon them in many ways. Meat was their main food during the glacial cold when plant food was scarce. From animal bones they made tools and weapons. They used tendons and intestines for cords. Hides and skins and furs gave them warm clothing, and walled in and lined their dwellings. Animal blood was consumed and used as dye for painting. These people drew the animals they hunted, as they saw them—horses, cattle, mammoths, bison, ibexes, bears, rhinos, muskoxen, reindeer, and so forth, as well as the predatory lions. The faithfulness of their reproductions has helped us picture some of the now-extinct animals that lived at that time.

The cave drawings, paintings, and engravings were not primitive scrawls. They were lifelike and inspiring, and have been proclaimed the finest animal reproductions ever done! They were not intended as household decoration, but must have had some magical or religious connotation, for they were hidden deep in caves, far back from the living quarters.

Ice Age animal art has been found in hundreds of caves and rock shelters in southwestern Europe, mostly in northwestern Spain and southern France, but also in Portugal, Italy, and places to the east. The world-famous caves of Altamira in northern Spain and Lascaux in west-central France are two of the more spectacular ones.

The first revelation came with the discovery in 1868 of the cave of Altamira, west of Santander. Its main ceiling and walls were decorated with pictures of red bulls and other animals, and adjoining galleries show bison, deer, goats, and boars done in many colors and wild confusion. Handprints and diagrams were also on the walls. The pictures were so magnificent, fresh, and modern in technique that they seemed recently made. Certainly they were not done by the "savages" of the Ice Age! They were dismissed as forgery or perhaps the work of Roman soldiers. But as more decorated caves and shelters were found and some already known were re-examined, the art's antiquity was proven.

Accidental discovery of Lascaux Cave in 1940 created another sensation. Its paintings are among the most beautiful and best-preserved examples yet revealed. Its vast oval chamber was painted to the ceiling with enormous bulls, deer, and horses, and other rooms were also arrayed with striking animal pictures. Tourists flocked to view this cave, one of the most famous sites of Paleolithic art.

Cave artists worked by the light of lamps fashioned of stone, and torches of branches or rolled birch bark, and small fires. Their pigments came from ochers, manganese ore, charcoal, and blood. They used various techniques—shading, dotting, watercolor effects, airbrushing—and carved in low relief. Many of their animals had a three-dimensional appearance.

The most fascinating thing about these pictures is that they were hidden in hard-to-reach, secret places—as though the artists were seeking the mysterious—where to see them one had to travel along hazardous, subterranean mazes; through narrow passages, underground rivers and ponds, and grottoes of dripping stalactites; down steep inclines; past whirlpools and waterfalls. To make some drawings the artists had to lie cramped in tight crannies. If the drawings were not openly displayed what was their purpose? Sympathetic magic? Were ceremonies held to ensure success in the hunt? Were adolescent boys brought to such caves for a soul-inspiring experience when initiated to hunter status? Archeologists feel there is a message in the drawings, diagrams, and handprints but it cannot be read.

As the last glaciation drew to a close and climate warmed, one would think culture would have taken another big step forward, but, oddly, it declined instead. This high level of art ended about 10,000 years ago, and would not be equaled in excellence until early historical times in Egypt, Mesopotamia, and China. It would not be reached again in Europe until the Middle Ages.

The Scandinavian ice sheet had been retreating for several thousand years from its farthest position across Germany, Poland, and Russia which it had held about 20,000 years ago and where it built prominent moraines. Deglaciation occurred more rapidly on the drier southeastern side than on the moister western side. There were several strong readvances but they were temporary and the ice sheet continued retreating. As it withdrew from continental Europe, it was fringed with a succession of water bodies, both ocean embayments and fresh-water lakes. Ultimately these became the Baltic Sea.

The flattening British ice cap divided into smaller, local ice caps and glaciers, and gradually disappeared.

Continental Europe and the Baltic Sea lowlands between it and Scandinavia were free of ice by about 11,000 years ago, as the ice sheet had by then pulled back to the peninsula. About 9,500 years ago the remaining ice sheet was thin and probably stagnant along much of its periphery. By 8,000 years ago lowland parts of the ice sheet had melted away and glaciers remained only in the mountains. The ice sheet was definitely gone by 7,000 years ago, and by 5,000 years ago most mountain glaciers too had disappeared or shrunken to small size. As the Ice Age ended, climate was even warmer than at present.

The return of mild climate brought more difficult times to central Europe. The woolly mammoth and the woolly rhinoceros finally became extinct. Horses, bison, and reindeer replaced them and were heavily relied upon for food and other essentials.

As the tundra moved north with the withdrawing ice for the last time, grass and shrubs took its place, followed by birch and coniferous trees, and then deciduous mixed forest. Game became less plentiful. There were still many animals about, but the big herds no longer roamed continental Europe. What was left of them migrated north. Among the last to go were the reindeer herds, which went north about 8,000 years ago. There they still live on the tundra but not with their former freedom.

About 9,000 years ago people of the Middle East and Egypt had learned to sow grain and domesticate goats, sheep, and other animals, and village life had begun. In southeastern Europe sheep, cattle, and pigs may have been domesticated by then too. Slowly knowledge of subsistence agriculture and animal husbandry, a whole new way of life, spread from those regions. It reached the Atlantic seaboard about 5,000 years ago, and England about 4,500 years ago. Forests were being cleared, and the game animals fled. Primitive hunters and gatherers either adopted the new agricultural methods or moved farther north.

As the Ice Age was ending in the Old World history was beginning there. By then the different races were evident in the world's population, but how they originated we cannot say.

Thus far we have told the story of the last glaciation and its people in the Old World. Now let us go back again in time and retell the story as it unfolded in the New World.

It is not known when or where Early People first arrived in the Americas. Dating of artifacts shows that big-game hunters were there during the height of the last glacial period, called the Wisconsin Stage in North America. (It is the counterpart of the Würm Stage in Europe.) But North America at least may have been thinly inhabited earlier, even before the Wisconsin Stage, which began about 75,000 years ago.

Recently found in eastern New York State are hundreds of tools of hard limestone, quartzite, and molded clay that seem to be over 70,000 years old, judging by the age of the soil that buried them. Their patina and weathering agree with such an age; and the flaked projectile points, hand-axes, scrapers, and knives are similar to Old World stone tools of that time.

Far less is known about Early People and their culture in America than in Europe, Africa, or Asia. Skeletal remains of Ice Age Americans are rare, so rare that it is thought cremation may have been customary. As far as we know, all who came were *Homo sapiens*, and all were Paleo-Indians (people developing toward the modern Indian) of one type or another, except for the Eskimos, who were among the last prehistoric immigrants.

The Wisconsin Stage in North America, like the Würm Stage in Europe, was a time of many climatic fluctuations, and hence of many fluctuations in the size and vigor of the glaciers, which affected migrations of people and animals.

The Laurentide ice sheet seems to have started forming from ice-accumulating centers in north-central and eastern Canada. As it grew it met ice shelves and glaciers spreading from Ellesmere, Baffin, and other islands between Canada and Greenland. Ice from those islands also met Greenland ice, which extended farther west than now.

Along the Atlantic coast, the Laurentide ice sheet pressed out past the present coastline onto the continental shelf, much of which was dry land due to lower sea level. Icebergs were discharged into the Atlantic in great numbers. The sheet reached south as far as New York City and Long Island. Its front curved west across central Pennsylvania, and dipped farther south in the interior lowland, but stopped short of the Driftless Area of southwestern Wisconsin and parts of neighboring states, that "window" through the drift that escaped glaciation many times while lobes of ice stopped north of it or bypassed it on either side. West of the Mississippi the ice front slanted northwest across the Great Plains and northern Montana.

The ice did not necessarily stand at all those extreme forward positions simultaneously. It would have advanced farther in different sectors at different times, and it waxed and waned.

The ice sheet grew thickest and moved most vigorously toward the south and east, for most of its moisture came from the Atlantic and the Gulf of Mexico. At maturity the sheet was easily a mile and a quarter thick, judging by the eastern mountains it overrode and comparing it to existing ice sheets. How much thicker it may have been we cannot say. It was considerably thinner and less active on its north and west sides.

In the Wisconsin Stage the ice did not come as far south as some earlier ice sheets had. They had reached to the Ohio and Missouri rivers, and beyond. They had gone to Kansas City, past St. Louis into southern Illinois, and into northern Kentucky. Now the ice bulged in lobes across the Canada–United States border, pushing farthest south through the central Great Lakes region into Illinois, Indiana, and Ohio. There the ice may have advanced at a rate averaging about 200 feet per year, according to radiocarbon dating of trees pushed over by the ice and of other organic matter buried in the drift. In some years it may have advanced 400 feet or more. Naturally the rate of advance varied greatly with time and location; and at times the ice retreated and readvanced.

The Great Lakes did not exist then, but the ice sheet's erosion helped create their basins. The basins of Lakes Superior, Michigan, and Huron are especially deep. Lake Superior's floor is 730 feet below sea level, Lake Huron's is 170 feet below, and Lake Michigan's is 340 feet below. The bedrock basins are even deeper than the present lake bottoms because they are partly filled with drift.

In western Canada the Laurentide ice sheet at times met glaciers from the Rockies. They were part of the Cordilleran glacier complex, a mass of coalesced glaciers which engulfed the snowy mountains of western Canada and coastal Alaska. They did not spread far east of the Rockies; their moisture supply came from the Pacific.

During the Wisconsin Stage the Laurentide ice sheet and Cordilleran glacier complex were at their maximum extent probably 20,-

ooo to 14,000 years ago. When and where they met at different places is not clear, and this is a subject of considerable interest because it bears upon the nomadic Ice Age people's freedom of movement in and through Canada.

It is generally assumed that the first people to reach North America came from Siberia. Though some people may have arrived during the last interglacial (by a route or routes unknown), the easiest avenue of entry apparently would have existed during the cold periods when glaciers were large and sea level was low. Then a land bridge emerged between Siberia and Alaska. By Late Pleistocene people had developed the means to live in the Siberian subarctic, so they were well prepared to enter Alaska even at the coldest periods of the Wisconsin Stage, which is when they could safely walk across the gap that is now Bering Strait. That is when and how most inmigrations are believed to have occurred.

Eastern Siberia, which today records some of the world's lowest winter temperatures, was intensely cold in Late Pleistocene, and glaciers covered its mountains. But the woolly mammoth and rhino were there, and the horse and steppe bison and other grazing animals. An exchange of animals took place between continents whenever sea level fell enough to expose the ocean floor. But it is not known just when the land bridge existed, in whole or part, and when it was submerged. It is likely that near the end of the Ice Age it existed around 45,000 years ago and again within the period from about 25,000 to 10,000 years ago.

At present Bering Strait is only about fifty miles wide, and the Diomede Islands lie midway between Alaska and Siberia. Even without a land bridge (as in this present non-glacial period) it would have been, and is, possible to cross the strait on pack ice in winter. There are no indications that Ice Age people of that area were seafaring, although they may have crossed small water bodies on rafts or primitive crafts.* It is doubted, however, that many families would have gone that dangerous, empty distance over ice or water without strong incentive. Or are curiosity and wanderlust incentive enough?

It is not known whether any Ice Age people reached America by crossing water bodies wider than the Bering Strait.

The Bering land bridge was not just a narrow isthmus. When sea

---

* People moving from island to island had apparently reached Australia over 30,000 years ago, indicating that ocean travel was possible then, in that area at least.

level was lowest, it was a broad plain up to 1,300 miles wide from north to south, unglaciated for the most part, carpeted with tundra and steppe vegetation, easily traversed, and well populated with animals from both Asia and North America. Hunters wandering onto it had no idea it was a link between continents. It was simply more living space and hunting ground, and they occupied it.

Bands of hunters who reached North America during the last glaciation undoubtedly roamed east up Alaska's broad, ice-free Yukon Valley into Canada. From there their descendants probably headed south through the corridor between the Laurentide ice sheet on the east and the Cordilleran glacier complex on the west. At some times the corridor was wide open and well stocked with game, but at other times it was pinched shut by ice in places, and so it controlled the passage of hunting bands to more southerly regions. It is unlikely that many Paleo-Indians traveled south along the Pacific coast, for that was a stormy region of high mountains and ice-scoured valleys from which glaciers sent icebergs crashing into the sea. The corridor route east of the Rockies led them onto the Great Plains, where they dispersed in all directions. They are known to have been at the southern tip of South America at least 11,000 years ago and very likely were there earlier.

Since America and Asia were periodically connected and separated a number of times, the migrations came in waves, and different groups brought different cultures.

In North America the Ice Age people hunted one of the favorite game animals, the woolly mammoth, which ranged south to north-central and northeastern United States. Other species of mammoth lived farther south. The hunters also found mastodons, the stocky, straight-tusked elephants of the forests which had become extinct in Eurasia; the large bison; muskoxen; the slow, stupid, elephant-sized ground sloth; several kinds of bears; camels larger than llamas; caribou, which are similar to reindeer; and giant beavers. Present too were antelope, elk, wolves, saber-tooth tigers and other large cats, foxes, rabbits, ground squirrels, and birds larger than eagles.

The West was less arid than now during much of the Late Ice Age, due either to increased precipitation or to lower evaporation. Much of what is now desert there was grassland then. Many western lakes that are now salty because they have no outlet were then larger fresh-water lakes, and fishing was another source of food. A number of volcanoes in the western mountains were active.

The broad continental shelf along the Atlantic coast emerged during low sea level, and was then a plain covered with tundra and coniferous trees and inhabited by mammoths and mastodons. Recently fishing boats trawling the ocean floor and searchers for sea-floor mineral deposits have come across some forty sites where bones and teeth of these animals were found. Some were as far as eighty miles off the coast. The teeth weighed up to eight pounds each.

Vegetation in North America was continually adjusting to changes of climate and the glaciers' size. During the climax of the Wisconsin Stage, tundra and forest-tundra fringed the ice sheet and extended south along the cold summits of the Appalachians. South of the tundra belt, open spruce forest seems to have covered most of eastern and central United States. Birch, tamarack, alder, and patches of deciduous trees were present too. Still farther south, trees of the boreal forest graded into temperate mixed forest. Open pine woodland grew on the southern Great Plains.

In the central plains strong winds blowing over bare, dry drift and floodplains resulted in great dust storms, sand dunes, and thick deposits of loess (wind-blown dust), some up to 100 feet thick. Loess also accumulated on the plateau of eastern Washington and Oregon.

Toward the end of the Ice Age the large animals decreased in numbers; the mammoth and bison and some other animals were a little smaller; and many species became extinct. North America saw the disappearance of its camels, horses, large bison, woodland muskoxen, giant beavers, ground sloths, mammoths, mastodons, and many other animals. Their extinction may have been caused by the repeated changes of habitat brought about by the ice's fluctuations, and by climatic reversals which interfered with regular cycles of birth and migration. Whether Man's hunting was a major or insignificant factor in the extinction of animals of North America and the world is a matter of debate among ecologists. Some think people had only a small influence on the balance of nature. Others think they overhunted and too often killed the young and female animals for their more tender meat. Still, many of the hunted animals did not become extinct, but merely dwindled in numbers and found refuges in undisturbed regions when Man took over their territory.

Though the mastodon is said to be one of many animals that be-came extinct thousands of years ago, Indians told early European settlers in America that their grandparents had seen such animals alive; and when President Thomas Jefferson sent Lewis and Clark to explore the Missouri and Columbia rivers in 1804, they were in-structed to keep their eyes open for mastodons. None was seen.

After a number of advances and retreats on its many fronts over tens of thousands of years, the Laurentide ice sheet began its final retreat about 18,000 to 15,000 years ago, beginning earlier in some sectors than others. In Ohio—one place where the ice's retreat can be gauged—the ice front drew back several hundred feet a year, and for a while over a thousand feet a year.

As the ice sheet withdrew* it also flattened and withered as an old snowcover does at the end of winter. The summits of long-buried hills reappeared like knees bent up in a pool of collapsing bubble-bath suds. Bare patches of earth, heated by the sun, quickly widened. The ice sheet became honeycombed with holes and tunnels. Its edges were no longer firm, but cracked and crumbling. In summer, when most of the melting occurred, sheets of water ran out from under the ice, and rivers gushed from its tunnels and poured off its surface —its lifeblood hemorrhaging away. The snowflakes that had been held captive for thousands of years were at long last released. Moraines of rocky debris were festooned along the ice-sheet margin, arced in the shape of the lobes that had pulled back. End moraines stood at the ice's farthest position, and recessional moraines were built behind at places where the lobes paused during retreat.

In front of the shrinking ice were outwash flats of braided streams where sands and gravels were sorted and spread out. Frequently chunks of dead ice were covered or surrounded by outwash, and later when they melted there were depressions, or pits, where they had been. Outwash with such depressions is *pitted outwash*. Similarly, ice blocks left in moraines were responsible for many of the *kettles*, or hollows, in those hummocky bands of hills.

Where the land sloped toward the ice sheet, meltwater collected in lakes along the ice margin. While ice lobes lay across the Great Lakes lowlands and St. Lawrence outlet, drainage was blocked and water that collected as lakes next to the ice took different exits at

---

* A description of the retreat of the North American ice sheet can be applied with appropriate variation to the Scandinavian ice sheet and other glaciers as well.

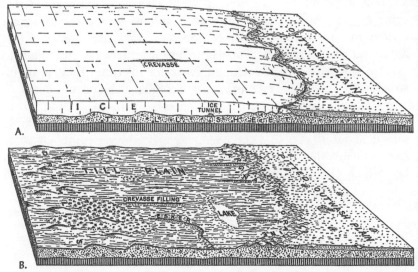

Diagram A shows a plain partly covered by the margin of a stagnant, melting ice sheet. Diagram B shows the same area after the ice has disappeared, and the various features left on the landscape. Notice how the esker formed in the ice tunnel. The pits on the outwash plain are caused by the melting away of chunks of ice buried in the outwash. (From *Fundamentals of Physical Geography*, by Trewartha, Robinson, and Hammond. Copyright 1968 by McGraw-Hill Book Company. Used with permission of McGraw-Hill Book Company.)

different times. So the levels of the lakes rose and fell, and the lakes assumed various shapes as the ice lobes moved back—or forward temporarily. The shape of the lobes can be seen at places in the outlines of the lakes. Lake Michigan is the best example; the lobe was larger than the present lake. The ice sheet was finally leaving the Great Lakes region, retreating into Canada, about 11,000 years ago.

Of the many other lakes along the margin of the ice sheet, Lake Agassiz was the largest. It extended from central Manitoba to central North Dakota. Lakes Winnipeg, Winnipegosis, and Manitoba are remnants of it.

At the southern end of the Cordilleran glacier complex, ice-dammed lakes burst out in one of the most catastrophic floods in the geologic record, creating the scabland of Washington with its network of spillway channels.

Altogether the melting of the North American glaciers released something like 8 million cubic miles of water. This happened over a period of about 10,000 years but, even so, the floods were tremendous and little water could soak into the still-frozen ground. Today's disastrous floods are trickles by comparison.

The innumerable streams pouring from the ice sheet joined and became concentrated in great *spillways*. Some of the main ones were the Mississippi (and numerous tributaries flowing into it), the Mohawk-Hudson River outlet across New York, and the Ottawa River in Canada. Much of the outwash flushed along the spillways was left as valley trains of sand and gravel in those river beds, for the amount of material was too great for even the mightiest of rivers to transport. Today the smaller rivers in those valleys are still slowly working on the islands and sandbars and fill, carrying the glacial outwash bit by bit downstream.

Sea level rose as glaciers around the world released their water—perhaps on the order of a few feet a century during periods of greatest melting.

As the Laurentide ice sheet continued wasting and thinning, parts of it became stagnant, and sections separated from the main body. Where it stood still and melted down without further movement, many delicate features it built beneath itself were left almost undisturbed. Among these are *eskers*, winding, snake-like ridges of gravels that were water-deposited in tunnels under the ice. They look much like abandoned, raised railroad beds trailing over the countryside, except that they sometimes go up rather steep grades, for the streams that made them were confined inside the ice and could flow uphill under pressure. *Crevasse fills*, which look a lot like eskers, were deposited along the base of crevasses into which water spilled.

As the ice sheet melted off the continent it left some areas scoured down to bedrock with no soil remaining. In some places rock was as highly polished as a marble tabletop. But most of the land was coated with drift. Plains and hill lands were generally left smoother than they had been, for drift filled old valleys; but moraines and uneven deposition of drift made some surfaces more uneven than they had been. Because former drainage channels had been dammed or buried, low spots contained many lakes, ponds, and marshes; and streams took circuitous routes and tumbled over rapids and falls as they picked out new channels. Glacial lake beds, when drained, were extremely flat areas.

The features the ice sheets built were not especially high. Moraines may rise 100 feet or occasionally 200 feet above the drift plain (although many are so low as to be hardly noticeable). They may be a single belt of hills if they mark one position of the ice, or they may be a band of hills a mile or several miles wide if they were built by a fluctuating ice front. In just certain places the ice sheet built *drumlins*, oval hills of gravel which may be 100 or more feet high and up to a mile long, though most are smaller. Where they occur they are often in groups, and all are parallel and aligned in the direction of the ice's movement. They are streamlined, and have the shape of a half egg cut lengthwise or, if elongated, of a cigar; and the end facing the direction from which the ice came is higher and steeper than the other end. It is not known why they formed as they did and where they did.

The ice left a forlorn wasteland, but not for long. Pioneer vegetation returned quickly and other plants followed as soil formed and the climate warmed. (At some places in Alaska where glaciers have retreated within historic time, bare drift has been reclothed with a succession of plants culminating in mature forest of large spruce within less than 200 years.) As the ice shriveled back, hunters with stone-tipped weapons moved into the drift-covered plains and lake country it had left.

The Laurentide sheet made several strong readvances during its retreat. The last major one—north of the Great Lakes—occurred about 8,300 years ago, while the sheet still covered a considerable area over and around Hudson Bay. Soon thereafter that last big ice body broke up and ceased to exist as a sheet. The breakup seems to have been caused by rising ocean waters coming toward it from the east. Around 8,000 years ago they leaked into the bay, causing calving and cutting into the ice sheet. The ice sheet divided into two main parts, both on high ground, east and northwest of the bay. Hudson Bay was nearly clear of ice by 7,000 years ago. The residual parts of the ice sheet are believed to have disappeared about 6,000 years ago.

Because thick ice remained longest over and alongside the bay, into the final phase of the Wisconsin Stage, that area was depressed longest and still is in the process of rebounding. Ice may have been about two miles thick there. When rebound is complete, Hudson

Bay will all but disappear, but the rate of rising now is only a fraction of an inch a year. Northeastern North America, like Scandinavia, has recovered about two thirds of its former elevation since unloading its weight of ice. Northern Britain and other areas that were heavily glaciated are also upwarping.

Around 6,000 to 7,000 years ago, it could be said, the Ice Age ended. Then the great ice sheets of North America and Europe had disappeared. By then the rapid rise of sea level had slowed. About that time the relatively stable period in which we live began. In Greenland and Antarctica the only remaining ice sheets are making a strong last stand. Unless the past pattern of ice-sheet withdrawal changes, it seems to be only a matter of time before they too disappear.

CHAPTER 11

THE HOLLOW SETTING

Within a few millennia our ice may be gone, but if so we shall have paid a fearful price for its going. Ice at its maximum extent is disastrous, but no ice is worse. The best stage is the halfway one, where we are this blessed moment, but like the others it too is unstable and will not stay put. We know where we are and which way we are going, but which way we shall be going a little later we do not know.

W. J. HUMPHREYS

SOMEWHERE in a scenic alpine setting is a lodge perched on a mountain slope overlooking a picturesque glaciated valley. Its spacious windows and wide verandas were designed to give its guests the best possible view of the magnificent glacier that decorated the valley opposite with its drape of glistening white. The glacier was the focal point of interest, the reason for the lodge's being where it was. Vacationers, health seekers, artists, nature lovers came from many countries for the inspiration that vista gave them.

From the lodge they looked across the valley to the energetic, billowing stream of ice that for thousands of years had been sculptor of this scene. In the clear mountain quietness they pondered the progress of time as exemplified by the slow-sliding glacier. They watched snow avalanching and ice falling from the precipitous rock walls in puffs of powdery spray and listened for the muffled thunder that followed. They lazily watched the glacier drop a few more rocks on its moraine and took delight in seeing its twinkling rivulets drain to the turquoise, iceberg-spangled lake below. The view from the lodge was painted by many an artist, detailed in many a letter sent home, and cherished in many memories; and the glacier's admirers returned often to study the changing scene, for the scene did change.

PLATE 3.· Glaciers have a powerful attraction that lures people to them in spite of death-defying dangers. Men venture onto the Bravo ice fall in the Coast Mountains of British Columbia, tiny specks on the treacherously fractured, tumbling glacier.

PLATE 4. Steep slopes do not deter glacier-climbing enthusiasts on the Mischbachferner.

PLATE 5. Ice caps on plateaus of Devon Island west of northern Greenland creep outward as ice caps must have done at the start of the Ice Age.

PLATE 6. When the last ice sheet moved across this part of Canada's Northwest Territories (from upper right to lower left) it grooved the land, leaving elongated, parallel gashes many of which are occupied by long, narrow lakes. The white patches on one lake are ice floes. The river is about ten miles from the bottom of the picture in this aerial view.

PLATE 7. A geologist studies one of the large glacial erratics in a moraine region of central Wisconsin (Waupaca County). Many of the boulders have been hauled to the fence lines in the slow process of clearing the fields. Most boulders here are granite.

PLATE 8. A broad view of Taylor Valley, Antarctica, showing various sizes of glaciers and how they expanded in many areas at the start of a glacial stage. Glaciers which were first confined to cirques spilled over and expanded downslope as piedmont glaciers, and when enough united a larger valley glacier formed. Here Taylor Glacier moves toward McMurdo Sound, but is unable to reach it as it did in the past.

PLATE 9. Aerial view of a cirque on 15,000-foot Mount Tyree in Antarctica. In the background a number of peaked horns can be seen. Valleys are deeply filled with ice and only the jagged summits of glaciated mountains are visible.

PLATE 10.   A moraine being built where debris collects from the melting, 150-foot-high front of Mueller Glacier, New Zealand. Tourists stand on part of the moraine built earlier. Note dark holes in the ice face from which boulders have fallen. In back is Mount Sefton (10,359 feet), which feeds this glacier.

PLATE 11. The snout of Meserve Glacier, Wright Valley, Antarctica, latitude 77° 35′ S, has a somewhat domed shape, being about 200 feet thick at the center and sloping to its cliff edges, which are about 65 feet high. Notice the difference in albedo (reflectability) between the bright white of the glacier and the dark, bare surroundings.

PLATE 12. Close-up of cliff of Meserve Glacier at its terminal point just before a large section of ice broke off and avalanched forward.

PLATE 13. Strong winds drift snow across Antarctica's Ross Ice Shelf into the sea. Parts of the ice shelf seen here are up to 150 feet above water. The shelf moves seaward, about 3 to 6 feet a day, fed by glaciers flowing off the continent and by snow accumulation. Its outer edge is broken off by violent ocean storms.

PLATE 14. Hole-in-the-Wall Glacier in Alaska knocking over trees as it moves ahead. Notice the upslope movement of the ice on the hillside.

PLATE 15. At Byrd Station, Antarctica, a thermal drill is sunk into the ice sheet to pull up cores of ice. Trenches such as this, cut in surface layers of glaciers in Antarctica and Greenland, serve as storage areas (notice stacked boxes of food) and contain well-furnished, prefabricated living quarters.

PLATE 16. Herds of shaggy woolly mammoths roamed the tundras and steppes of Ice Age Eurasia and North America. They probably used their impressively long, curved tusks to clear snow away from the low vegetation which they ate.

PLATE 17. A Neanderthal man burns the points of wooden spears to increase their effectiveness. At upper right a temporary shelter is made of poles and animal skins.

PLATE 18. Artists of the Late Ice Age at work in a shadowy cave. Torches and oil lamps gave light. Crude earth materials ground up provided colors for the paints.

PLATE 19. A thick deposit of loess near St. Louis. Loess forms bluffs along the Mississippi River, which was a major spillway of the wasting ice sheet, where accumulations of wind-blown dust were heavy for a long time.

PLATE 20. A glacier cave in Victoria Glacier, Banff National Park, Canada. Eskers were often deposited in such tunnels.

Today the lodge stands vacant. No guests lounge on its verandas. Its wide windows are blind. No one looks at the glacier any more because the glacier is gone. The valley now is just another empty trough. At first the glacier had backed up only slightly. Then it began to be not quite so attractive. It lost its healthy whiteness and flattened like a deflating flabby balloon. And then it retired, year by year, farther up the valley. Finally just a glimpse of it remained at the valley's bend, and then it withdrew from sight as an actor exits into the wings.

The site described is no specific one. It is a generalization of many such sites where a magnificent view is no more, for the ice—the jewel that gave sparkle to the setting—now is dull and dirty and dying, or has disappeared completely, leaving only a hollow setting.

Every year some glaciers die. Their passing has little immediate significance or news value, so no obituaries are written except in concerned journals such as *Ice* or the *Journal of Glaciology*. The habitat of most glaciers is off the beaten path. Myriads of them have not even been named. Some will leave this world without having been recorded on a map, as thousands already have done. Some that are inked on maps have already vanished. Some will have been seen only by fliers and some seen never at all.

Many a glacier now on the verge of extinction is only a scab sticking to a mountainside or a dab of leftover pudding in a cirque bowl overlooking the deep, intricately carved valley it hewed when in its virile years.

There is not only a general lack of concern about the shrinking of many of our glaciers, but a nonchalant attitude of good riddance and let's hurry it along. The common attitude that ice is superfluous, expendable, and in the way makes the danger of its loss all the more plausible. Much of the world's ice is in danger of being lost, not through nature's attack alone, but through nature aided ably by people, as we shall see in coming chapters.

All glaciers may disappear sometime in the future—when the world warms sufficiently again; but this will not happen in our time; so it would seem the prospect is hardly worth wrinkling a forehead about. In some people's minds there is so much ice around that one should think not how to protect it but how to dispose of it.

Glaciers are worth protecting not only because of their esthetic qualities and their link with the prehistoric past, but for their role in

storing water, evening the flow of rivers, and regulating climate, for
while they are influenced by climate they in turn have an influence
upon the climate of their regions and of the whole world. Also they
are some of our last genuinely natural, virgin, wild areas. Take the
ice away—even just a sizable fraction of it—and this would be a
noticeably different Earth with geographical conditions perhaps less
hospitable to humans than those present ones in which we thrive.

Nothing is so large, so firmly fixed upon this planet, that it cannot
be ruined or erased. It is not necessary that every last smidgen of a
thing be gone before it is destroyed. You might have a heap of
crumbs but not a cake, or a scattering of trees but no true forest.

Thick, continuous forest used to cover all of eastern United States
and southeastern Canada. The first colonists in the East thought that
obstacle to settlement would never be conquered. How to get rid of
all those trees? They chopped them down. That was too slow. They
burned them to make clearings. Yard by yard, hill by hill the forest
was removed. Population increased, the rate of cutting and burning
accelerated, and soon, in only a few generations (by 1830 in the East,
by 1900 in the north-central area), before it was realized what had
been lost, the virgin forest was ruined—all its irreplaceable wealth of
high-grade timber, of water-storing and soil-retention ability, of
healthful solitude, wildlife, and sylvan beauty. You still see many
trees in the deforested area—poorer grades of second growth in
patchy woodlots, clumps of shaggy juveniles, crippled nondescripts
lined up with telephone posts. But the primeval forest of stately
monarchs growing close and uniform, that green-ceiling cathedral
spreading its shade almost halfway across the continent, disappeared
in an unbelievably short time.

Beyond the forests lay the prairies. To the pioneers crossing them
the man-high grasses, the profusion of colored flowers, seemed another endless, wasted realm that had to be torn up, mastered, and
put to use. Fires blackened the plains to tame it. Cattle were turned
loose to trample and graze upon it. Finally steel plows ripped up the
mat of roots and alien plants were substituted. Now it is almost impossible to find one acre of original prairie plants. Years ago Aldo
Leopold in A Sand County Almanac wrote regretfully: "No living man
will see again the long-grass prairies, where a sea of prairie flowers
lapped at the stirrups of the pioneers. We shall do well to find a forty

here and there on which the prairie plants can be kept alive as species."

Glaciers cannot be kept alive as little sprigs and fibrous roots here and there as plants can be. They cannot be hatched in climatrons or bred in zoos. They need space and freedom and privacy unlimited in which to breed and flourish.

No one has yet found the answer to the most fundamental question about them: how do they move? We are not even close to knowing. What marvel of mechanics propels them? Their power of motion is not simply a matter of sliding downhill, because, although gravity helps their movement, they can move on flat land and up slopes and can travel hundreds of miles over any terrain, so some internal action gives them their drive. Is it slippage along the planes of ice crystals, or the rotation and reorientation of their crystals, or something else?

Investigators with several university degrees and years of field experience seek the answers without success. They drive stakes into glaciers to measure their movements, bore into them, drill thousands of feet down to their very base, take soundings and series of photographs. They camp on them, live inside them, write reams of reports and equations about them, and still cannot fathom the mystery of their movement.

Other of nature's creations can be tested in laboratories or duplicated in miniature, or be dissected, or reproduced in controlled environments to see how they operate. Soil and rock can be sorted and handled. Water and air can be seen through and moved through. But glaciers cannot be reduced to toy size, duplicated in mockup, or held captive for domestication. Probe them. Smash them up. Melt them if you will. Carve off thin slices of their ice, shoot all kinds of light through them, and magnify them. Drill ice cores out of them and pick them apart. But still their secrets remain. Naked and exposed they present themselves, but still their blank mask and opaque body give no answer to why they move or what they will do next.

Old the glaciers are, but old age is not without strength and beauty. It has a steadfastness and patient composure. Those glaciers which were old when our civilization was aborning, those silvery shocks on the mountain brows, give a grandparental richness and sedateness to the landscape that nothing else can.

If anything symbolizes timelessness, glaciers do. So it seems strange for anyone to deplore the dwindling of our glacier assets. But for the present let us think of just those accessible glaciers with which we realistically can have contact. After all, not many people plan to visit Greenland or Antarctica or even the heavily glacierized Arctic islands. Omit those for the time being and we face a different situation entirely. We are left with a thinly distributed scattering of quite vulnerable caps and streams and patches of ice.

If you would see glaciers, see them now. That sounds like an exaggerated admonition. But don't believe for a moment that many of our grandest glaciers are not in danger of being ruined within a fraction of a lifetime. If there are glaciers you have wanted to see, do not wait too long. The years may not deal kindly with them. Because glaciers have such long lives and have been around at least as long as the human race, they are taken for granted and expected to remain forever.

Perhaps there is a particular glacier you have seen and like to remember. You may have camped or lodged where you enjoyed the sight of its graceful form brightening the mountains from which it cascaded like a frozen waterfall, and watched the rosy alpenglow tint its heights ethereal pink in the sunset and dawn. You saw it glowing spooky blue in purple twilight as though it held a faint inner light, and gazed at its silent mysticism under a full moon.

Next time you see it you may be shocked. Chances are the glacier will be smaller and less spectacular than when you first saw it. Consider yourself fortunate if it is not.

Back in 1872 John Tyndall, British professor of natural history, mountaineer, and glacier enthusiast, began the preface of one of his books with these remarks about glaciers of the Alps:

After an absence of twelve years, I visited the Mer de Glace last June. It exhibited in a striking degree that excess of consumption over supply, which, if continued, will eventually reduce the Swiss glaciers to the mere spectres of their former selves. When I first saw the Mer de Glace its ice-cliffs towered over Les Mottets, and an arm of the Arveiron, issuing from the cliffs, plunged as a powerful cascade down the rocks. The ice has now shrunk far behind them. . . . The vault of the Arveiron has dwindled considerably. . . . The ice-cascade of the Géant has suffered much from the general waste. Its crevasses are still wild, but the ice-cliffs and séracs of former days are but poorly represented to-day. . . . The great Aletsch and its neighbors exhibit similar evidences of diminution.

Literature is full of such expressions of regret concerning glacier regression.

A friend of mine, when he was a boy, used to play at the foot of a mountain glacier in western United States. Now middle-aged, he recently returned to his old home. He was amazed to see that the glacier's terminus is halfway up the mountainside. It is too high to reach now without a strenuous climb. Photo files in glacier research centers are full of before-and-after pictures of glaciers showing their retreat and their change from beauty to ugliness. That glacier you remember may, in not too many years, have wilted like a flower in the sun.

Sometime you may have boated up to the cliff-high terminus of a gorgeous glacier that met its own reflection in a lake or sea, and there you waited in risky suspense for the bouncing waves to jar teetering chunks off its cracking front. You watched sharp-angled icebergs plunge into the water, submerge, and then rebound into the air spraying fountains of water about them in christening showers for their launching while churned waters threatened to capsize your boat. And then you watched the icebergs settle down like glass castles drifting majestically and shimmering blue-crystal in the sun.

Go back again years later and the icebergs that thunder to the water may be fewer and look like miniatures. The glacier now may not even reach down to the water's edge. Its high cliff front may be squashed and grounded in moraines as black as the tailings from a mine. It may be so covered over with its own refuse that it is totally invisible. It may not be there at all.

The shrinking of glaciers, thousands of them, causes little fuss. That has been going on as long as we can remember. We feel no radical, world-shaking change, nothing to be scared about. Contrariwise, the retreat of glaciers gives a feeling of relief, of safety and release from what we think of as a scourge of the past.

But if we were to tell of glaciers advancing, that is a different story. Listeners perk up. One glacier surging forward faster than usual makes headlines. Glaciologists dash to see it. It is placed under surveillance. Definitely, those who speak of the possible regrowth of glaciers have a more eager audience than those who speak of the danger of losing them, because it is more exciting to imagine the Ice Age returning than to think of climate becoming milder.

Not all glaciers worldwide behave the same. Some are advancing while others retreat. During the long course of retreat there are times when the ice, like an old performer making a temporary comeback, has staged spectacular readvances in certain areas. It could happen that the ice, right now, might make another grand entrance, not just locally but all over the world, and our talk of recession would be outdated.

The alpine glacier we spoke of earlier, which retreated out of sight of the lodge—is its retreat only temporary? What if a decade or two from now it should show itself again, rejuvenated and on the move, coming around the corner of that valley's bend? Would it not be welcome and would not the valley regain its old charm? The lodge could once again become alive, and visitors could again have their spirits lifted by its presence.

What if the ice kept coming? And coming? What if it rose thicker in the valley and pushed its cold front right toward the lodge's doorstep? Would it then be so welcome? Would not delight in its return change to alarm and a wish for its retreat once more?

The changes taking place in glaciers in our lifetime probably will be unfelt by most of us. And it will be a long, long time before the ice is all gone or back in force, perhaps to threaten our civilization. We are not inventing a false scare situation or pushing a panic button. We are just saying, let's watch what is going on and be aware of it. What is the situation regarding recession and growth of glaciers? What is going to happen to them, and is that what we want to have happen?

CHAPTER 12

SIPPING FROM THE TEMPTING CHALICES

. . . What an unspeakable luxury it was to slake that thirst with the pure and limpid ice-water of the glacier! Down the sides of every great rib of ice poured limpid rills in gutters carved by their own attrition; better still, wherever a rock had lain, there was now a bowl-shaped hole, with smooth white sides and bottom of ice, and this bowl was brimming with water of such absolute clearness that the careless observer would not see it at all, but would think the bowl was empty. These fountains had such an alluring look that I often stretched myself out when I was not thirsty and dipped my face in and drank till my teeth ached. Everywhere among the Swiss mountains we had at hand the blessing—not to be found in Europe, *except* in the mountains,—of water capable of quenching thirst.

MARK TWAIN

Perhaps in the long run beauty and inspiration are fully as important aspects of this environment, as a habitat for man, as are food and energy. . . . Perhaps if we sacrifice even such a small thing as a species of bird, or such a large thing as the Greenland ice cap we may find that subsequently the world will not so well satisfy our needs.

COMMITTEE ON APPLIED ECOLOGY,
ECOLOGICAL SOCIETY OF AMERICA

IN THE period of history called ancient, and undoubtedly long before that, people of arid central and western Asia were already tapping the snow and ice of the highlands to keep streams running in the foothills and lowlands. It was elementary for a sower of seeds to notice how water flowed from melted snow and how a scrape of a stick could channel it a few feet to a gully that drained to a planted plot. Rudimentary irrigation. Later, using diversion on a larger scale, water could be channeled in from a source farther away.

As a nomad sat idly alongside his goats and sheep which grazed on the choicest green sproutings, those in moist soil next to melting snowdrifts, he noticed that water trickled most from darkest snow, and stones that warmed in the sunshine melted the snow around and under them and sank into holes they produced.

Then sometimes when the snow kept its moisture locked up too long these observant, experimental agriculturalists knew how to release it. They could *make* it melt. They sprinkled earth and stones on it and awaited the desired results.

Today these arid parts of Asia must support many more farmers and herders. Most of the people there still make their living directly off the land. In earlier times that was a good life. Now it is poverty. The population pressure, the need to make one's land produce, becomes more acute every day.

One cause of aridity in many water-deficient regions is mountain barriers which create "rainshadows." But these same barriers that produce deserts or near-deserts often provide water to alleviate the dryness for which they are responsible. What they took away they in part return. On their summits they collect snow, some of which may become glaciers, and during dry seasons meltwater is rationed in streams to parched lowlands.

Today even in humid lands water is precious. After easily obtained supplies are all in use or contaminated, planners look around for new supplies. They see rivers which keep bringing water. If a river's volume fluctuates the question sooner or later arises whether the flow could be increased in low-volume periods. Tracing tributaries upstream, the planners are in many instances led to high snowfields and glaciers whose melting rate controls the volume to a large extent.

Are glaciers, which seem securely frozen assets, in danger of being liquidated? Although ambitions to exploit the tempting glaciers may be still small-scale or half daydream or little publicized, the course of planetary events will cause those ambitions to grow stronger.

This is no longer just a natural world. It is becoming more artificial all the time, modified by people's desires and needs. People can engineer unprecedented marvels—and unbelievable botches. Their attempts are still exploratory (and therefore especially dangerous), but their involvement in nature interference grows. Can they interfere with the life cycle of a glacier? They can. Would they ever try to hurry the melting of a glacier? They would and they do!

Let us look more closely at Inner Asia. Dry. The heart of the world's largest land body, far from oceans, walled off and compartmented by mountain ranges, including the highest on earth. But an inviting region because new productive land is scarce, and much idle land there could be producing if only water were available. Western China and the southern part of Asiatic USSR lie within this dry area.

For uncounted centuries people of Inner Asia who lived beneath ice-capped mountain ranges have been nipping at the water reserves of the glaciers.

Even before China's population pressure was as explosive as it is today the watered parts of its dry interior lands were settled. Oasis towns and caravansaries in foothills where streams descended from the mountains were stopping points for camel caravans traveling from China to India, Asia Minor, and the Mediterranean. During droughts the peasants climbed the mountains and spread ashes and dirt on the ice and snow. Sunlight is strong in thin, clear mountain air. The dark particles absorbed solar energy, radiated heat, and the snow and ice melted faster than they otherwise would have. Each speck became a little heater, melting itself into a pit, piercing the armor of the ice. Streams ran fuller and fresh water came into the fertile alluvial fans and river valleys below.

The practice of forcing snow and glaciers to melt was not an uncommon one throughout arid parts of Asia. It continues today and is spreading.

China's population now is one fourth of the world's total and its rate of increase is one of the highest. Its dry marginal lands must be put to more intensive use. The Tarim Basin of Sinkiang Province is one of the areas where China's agricultural expansion is being pushed. With irrigation water its land can produce grains, cotton, livestock, and luxuriant gardens with pears, apricots, grapes, and melons. It has great potential but its agriculture depends upon streams from the glaciered mountains that surround it.

Along the northern side of that oval basin rises the Tien Shan, a mountain chain reaching up to about 24,000 feet. (By comparison, North America's highest point, Mount McKinley in Alaska, is 20,320 feet, and Europe's highest, Mount Elbrus in the Caucasus, is 18,481 feet. The highest point in the Alps is Mont Blanc, 15,781 feet.) The Tien Shan's glaciers cover an area about four and a half times that of the glaciers of the Alps. On the western side of the basin lie the many-glaciered Pamirs, a lofty knot of mountains called "The Roof of the World," which reaches over 25,000 feet; and the Karakoram Range exceeding 28,000 feet. In these rocky towers are some of Asia's grandest glaciers, some of the most robust outside the nearby Himalayas, where Mount Everest, the world's highest peak, tops them all at 29,028 feet. On the south of the Tarim Basin is the Kunlun Shan, along the Tibetan Plateau. On the east is the Nan Shan, another high range with numerous glaciers, and a low opening which is the entrance to the Gobi Desert.

Over the centuries the amount of water supplied by mountain glaciers to the Tarim Basin has been gradually failing. Some oases that were prosperous are now uninhabitable and some former streams no longer reach the mountain base. This diminishing water supply is not in keeping with plans of the Chinese government. It wants more irrigation water and wants water earlier in the spring to lengthen the growing season, increase yields, and permit a greater variety of crops and more livestock, so it is trying out methods of inducing greater flow and forcing glaciers to deliver earlier.

Glacier-melting experiments are carried on with success in other areas as well. Fluctuations in the size of many Asian glaciers are being recorded because of the future importance of these ice reserves. Retreat of glaciers seems to be the general trend in central Asia. Yet increased reliance on artificial melting to support growing populations in marginal areas is almost sure to come. This could imperil the existence of many glaciers.

The rugged mountain systems of Inner Asia are not only divides between drainage basins but also buffer zones and boundaries between rival peoples. They separate China, the USSR, India, Pakistan, and Afghanistan, all of which have an urgent need to increase agricultural production and a growing thirst for water. The mountains protect and water little Nepal, Sikkim, and Bhutan also, and help fill rivers vital to Southeast Asia. With the world's highest peaks and Asia's largest glaciers apportioning water to rivers that flow outward in all directions to most of the countries of the Far East, we can anticipate how more than one nation may eventually be coveting the same glaciers and ice caps. We need only remember the bitter struggles among settlers of America's West for its waters, or note the heated campaigns among western states even today for water rights to certain mountain-fed rivers. Already we find both China and Russia engaged in glacier melting in mountains along their mutual border.

Almost all farming in the desert regions of Inner Asia is based on irrigation with glacier meltwater!

Who will control the glaciers of a war-prone, hungry Asia? As yet the impenetrable ice-capped mountains are not accurately surveyed or mapped, and the location of boundaries and crestlines is indefinite. Someday they will be in the limelight. There may be battling to get or to protect the now mystically remote and untouchable Asian glaciers. (Certain of them are considered holy. Kanchenjunga, the world's third-highest peak, is sacred to Buddhists, and the government of Sikkim will permit no one to go to its summit. Hindus make difficult pilgrimages over high passes to a glacier in Kashmir which they worship.)

Places that used to be totally inaccessible are now within reach from the air if not from the ground.

Hop across the glaciered mountains that separate China's Tarim Basin from Pakistan. On the other side is one of the world's largest irrigation projects. Pakistan—part of it Thar Desert, all of it dry—survives on irrigation water and looks to the mountains, the Karakoram and Himalayas, in whose white peaks the Indus River, her lifeline, begins.

Next door to her is India, where famine and drought constitute a perennial way of life. India relies for much of its agricultural production on the plains of the Ganges River, which flows west-east along the foot of the Himalayas and whose highest and most dependable tributaries begin in the glaciers of those mountains.

Destitute people who own little, who live from day to day in lethargic poor health, who are fated to die young, do not worry about the future's resources as do healthy people with amassed belongings, investments, and long life. The only important thing is to live till tomorrow, or if lucky till next year.

When one year of drought follows another, when the last hoarded rice is eaten, when your children die and the brown rivers dry up and the hot plains crack, and the skinny, sacred cattle nibbling on the land eat better than you do, you may well lift your dark, sunken eyes to the misty, godlike mountains and wish them to send their moisture for you and your crops, and to wash the stinking rivers clean. But the mountains hold high their cool, brimming chalices of ice water, offering just tantalizing sips.

What do a government's leaders do when the monsoon fails and its people gasp for help? Trucks and carts haul water to the parched villages. Human skeletons limp into cities from the foodless, waterless countryside. Malnutrition invites disease, epidemics flare, and people die in the streets. Mobs panic in hunger riots. Recent famine years saw trains wrecked in India by hungry protesters, business and communications disrupted, schools closed in some areas. Thousands of wild elephants, also drought victims, rampaged in five states. They trampled farms and ate crops, and near Darjeeling they stopped trains to get the grain and sugar cane, drank water supplies, and forced villages to be evacuated.

When mass meetings are praying for water, when the storehouses and reservoirs are empty, what does a government do? Would it not think seriously of turning the faucet on in the mountains, if that were possible, to avoid some suffering and a political crisis? That radical solution is not being advocated here but in some quarters it is being thought about. (It could be a dangerous experiment because if too much water were released floods might result.)

The USSR, while not that desperate, can also use more water from central Asia's glaciers, for she is pushing development of her marginal lands. She has plenty of cold, damp land with a short growing season. What she craves is farmland in a warm climate, and such land within her boundaries is limited. But east of the Caspian Sea lies a subtropical "promised land," her southernmost territory in Asia. Most of it now is scrub desert, but its soil fertility, long growing season, and abundant sunshine make the area too ideal for agriculture to be neglected just for lack of water. Water will be provided somehow.

Most streams dry up when they enter this region but it contains a few perennial rivers along whose courses there are narrow oases where cotton, sugar, rice, and tropical fruits are grown. The largest of these rivers are the Syr Darya and the Amu Darya. They are navigable in part of their courses and flow into the salty, shallow lake called the Aral Sea.

All agriculture in this trans-Caspian area has to be irrigated, and the principal source of water is some of the same glaciered mountain ranges that feed rivers running into China and the hungry countries to the south. The main rivers of this arid Asian section of the USSR start in the Tien Shan and Pamirs (which also contain headwaters of rivers flowing into China's Tarim Basin and Pakistan's Indus Valley) and in Afghanistan's Hindu Kush Range. There is an urgency to take action because the volume of water in the lower courses of the USSR's important central-Asian rivers is falling off. Large quantities are being taken out at the mountains for irrigation. Navigation is suffering as well.

For good reason the USSR shows an escalating interest in glaciers. Glaciological research has increased rapidly in recent decades, especially in the mountain regions of Inner Asia.

The Tien Shan and the adjoining Pamirs are the largest area of glaciers in the USSR. In the Pamirs is the USSR's largest valley glacier, the Fedchenko.

In the book *Soviet Geography: Accomplishments and Tasks*, published several years ago by the Geographical Society of the USSR, we find this ominous statement: "The industrial conquest of certain desert regions and the development of animal husbandry have required searches for new sources of water supply under the difficult conditions of an arid climate. . . . Hydrologists must look ahead and possess a 'reserve fund,' the liquidation of which may be required at any time by practical demands."

The opinion that glaciers will be required to supply more irrigation water in the future has been stated openly in scholarly papers and elsewhere, and much has been written about the dusting of glaciers to promote melting. This practice has been opposed by some glaciologists of the USSR who point out that artificial intensification of melting may "disturb the natural evolution of glaciers and reduce normal stream flow," but they are drowned out by others calling for the practical application of glacier-melting techniques.

Soviet scientists did not start out blindly in their glacier-melting programs. For years they had been testing techniques for melting snow and ice in the lowlands of Europe and Siberia by spreading ashes, slag, soil, coal, and soot. One snow-dusting experiment on the north coast of Siberia showed that dusted snow melted eight times faster than undusted snow; and another test made on an airfield on the middle Volga was equally enlightening. There the snow to begin with was 41 centimeters (16 inches) deep. It was covered with ashes on March 18, and was completely melted by April 7. Untreated snow was still 11 centimeters deep on April 19.

Artificial-melt procedures were used to open icebound rivers and ports of the north. Dust was spread over ice jams by low-flying airplanes.

Attempts were made to open the White Sea approaches to the port of Archangel (64° N) near the Arctic Circle. A mixture of dust and rock salt was spread over the ice, and then waste oil was poured on. Ice treated in that way broke up about fifteen to twenty days before untreated ice. The cost was estimated to be only 2 per cent of the cost of using icebreakers to open the channel.

Some Soviet hydrologists favor quickie methods of getting more water from the mountains. One method is producing artificial avalanches, shocking the snow and ice into falling downslope to lower elevations where it will melt more quickly. Artificial avalanches are being produced in mountains west of the Pamirs in central Asia, and similar avalanche stations are being set up in the Pamirs too. Another method is siphoning water from glacial lakes where meltwater has collected. Even if these methods did not touch glaciers directly they could hurt them, for snow and the lakes' ice water hold down temperatures in the glaciers' environment, helping to make their existence possible. When snow and the cold water are removed there is bare rock in their place, which heats quickly in sunlight and then radiates heat.

In USSR tests in the Tien Shan, coal dust and loess soil were spread by hand over glacier surfaces. Coal dust, being blacker, worked better. Only a thin dusting is required. (A thick layer would actually protect the ice from the sun. A covering as thin as three eighths of an inch, or about a centimeter, can keep underlying snow or ice from melting.) Five to ten tons of coal dust were applied per square kilometer (about four tenths of a square mile). The dusting reduced the albedo of fresh snow or firn by 25 to 30 per cent, and of glacier ice by 7 to 10 per cent. Dusting of entire glaciers, it was determined, could increase the annual stream flow by up to 55 per cent. The tests proved beyond any doubt that glacier dusting substantially increases the flow of mountain streams.

Next came plans to dust the Tien Shan glaciers, not by hand, but by airplanes. And not just one glacier at a time. One experiment alone dusted nineteen glaciers. Runoff increased markedly.

Glaciers were also blasted! As one report stated: "In addition to dusting of glacier surfaces, the program also included blasting for mechanical and thermal destruction of the ice." Canals were cut in the ice too and runoff was further increased in that manner.

The attack on glaciers has begun.

One can understand glaciers being melted by governments obsessed with production quotas, or water-short people at death's threshold, or backward countries that do not use conservation practices. But would a rich, conservation-minded, nature-loving country think of doing that? Well, glacier-melting experiments are done even in the United States.

Sipping from the tempting glacial goblets has started, and those who sample that nectar will be licking their lips in anticipation of larger and larger draughts. Others will want a taste too. And after the goblets are empty, who can refill them?

In South America hydroelectric plants in the Andes have increased their output substantially by using black dust to darken the glaciers whose meltwater turns their turbines. Results are called "encouraging" and plans have been made to expand the dusting operations.

Norway stepped up its glacier research recently at the request of water power authorities. Studies were made on glaciers located in catchment areas supplying water to hydroelectric plants to record how much the glaciers are fluctuating. That country, which was once the birthplace of Europe's greatest ice sheets and whose coast is lacy with ice-designed fiords and islands, now has only a scattering of glaciers left and most of them are small and shrinking. Due to a growing concern about the water supply, the Norsk Polarinstitutt and the Norwegian Water Power and Electricity Board have remapped many of the glaciers and are intensively studying them. The largest remaining ice cap in all of Europe is the Jostedalsbreen in southwestern Norway, just north of the famous Sogne Fiord. It is one of the most closely watched areas, partly because of hydroelectric power facilities near it.

Mountainous countries naturally rely heavily on hydroelectric power. It is the cheapest kind for them and sometimes the only kind. They have the requirements: steep drops in river courses, and abundant, reliable precipitation. Where much of the precipitation is snow, glaciers may exist.

In the Alps the principal kind of power is, therefore, hydroelectric.

The mountains lack mineral fuels, but they have fast-flowing rivers fed by ever-present glaciers and snowfields, and numerous glacier-created lakes serving as reservoirs to regulate water flow. Electricity provides the power source for almost everything, even railroads, and it is distributed to neighboring regions.

The Alps are also a water-distribution hub in one of the world's most densely populated regions. Their glaciers are the starting points of many of Europe's rivers, including the Rhine, Rhone, Po, and Danube. The Alps belong to Switzerland and Austria, and to northern Italy, eastern France, and northern Yugoslavia as well. With so many countries relying upon this compact area of glaciers—not only those countries that contain glaciers, but all those that use the river water and electric power that flow from there—and with all the commercial and political antagonisms among the European countries, how long can the Alps' glaciers remain unmolested?

As early as 1765, H. B. de Saussure, a Swiss physicist and alpine traveler, wrote that in the Chamonix Valley near Mont Blanc farmers scattered dark material on their snowfields so they could begin field cultivation two to three weeks earlier than usual. Many present-day inhabitants of the Alps may have reason to use this long-practiced technique to melt glaciers.

Local climate in the Alps varies from one valley to the next. In mountains cold breezes blow down the slopes and valleys, and warm air rises in convectional updrafts. Those valleys that do not have glaciers in their upper ends experience warmer weather in their lower, habitable parts. There crops can be started earlier and are not hurt by the strong, cold winds which drain down those valleys having glaciers at the head. The cold glacial air can whip down with force for miles beyond the glacier, and even at the mouth of the valley the cold wind is felt and suppresses the warm air that ordinarily would enter. These winds damage vegetation by blasting it with sand and snow. They depress the upper limit of possible grass growth as much as 1,500 feet, thereby reducing the extent of pasture, a resource valuable to the dairy industry which is important in the mountains. The growing season, already short at high elevations, is even shorter in valleys with glaciers at the upper end. It could be lengthened ten to twenty days near the mouth of glacier valleys if the cold downdrafts could be eliminated.

In time might there be pressure from local inhabitants to have more glacier-free valleys? Might farmers in a refrigerated valley like to see their crops do better and campaign for removal of the glacier? Might they even try to cause its shrinkage or disappearance themselves? Actually, small-scale attempts have been made to artificially melt glaciers in the Alps. Because in a half century the area of the Alps glaciers has decreased 25 per cent through natural wastage, there would be reason for concern if artificial melting were to be undertaken in earnest while natural wastage continued.

Let's look somewhere else. Canada is the second-largest country in the cold latitudes, after the USSR. Like the USSR, it has an annual battle with snow, and its lakes, rivers, and ports are choked by ice much of the year, so it has diligently been experimenting with artificial melting of snow and ice. This is the stage of attack the USSR went through just prior to its direct attack on glaciers. Will Canada follow the same course? It could be tempted to, for it too has a water-deficient region which is rapidly being settled and promises to add great wealth to the nation's economy if it can be developed. And nearby stand mountains heavily coated with snow and ice.

In the Coast Mountains along the windward Pacific side Canada has some of the finest glaciers outside polar zones. Some are in British Columbia and some she shares with Alaska's panhandle. In the Rocky Mountains farther inland are lesser, but still impressive, glaciers and a valuable snow supply. Among those mountains are many small basins that could become more productive agriculturally. And on the prairie east of the mountains is the big boom area of western Canada. It lies in the rainshadow of those high snowy mountains and is semi-arid grassland. That natural vegetation develops dark steppe and chernozem soils, the most fertile and arable of all residual soils. Part of the land is old glacial lake bottom, also fertile and flat, easily cultivated or built upon. What has been cattle and grain country is being used for more intensive, higher-paying irrigation agriculture. Land that has been idle is being cultivated. Oil has been discovered. Industry is moving in. Cities and towns are growing fast. Water is needed in quantity for the rapidly expanding municipalities, for mining and other industrial operations, and for the spread and intensification of agriculture. Canada's leading port is no longer Montreal in the east, but Vancouver on the west coast. The critical time has not yet been reached, but as Canada builds westward into the dry lands the mountains with their snowfields and icefields may be called upon to deliver more water.

There will be other calls for water from those mountains too. Water is a highly salable resource. Just noting that the Columbia River, which is vital to northwestern United States, heads in Canada's Columbia Icefield and in other nearby glaciered mountains hints of future possibilities and pressures. That river is one of the world's greatest dispensers of irrigation water and producers of hydroelectric power. United States' population is shifting west also.

Canada is quite experienced in techniques of melting snow and ice. Dusting is commonly used to keep waters open to navigation and to speed spring break-up not only in the Yukon and Northwest territories but in southern provinces as well. It is not practical the whole winter; it is most effective in the normal melting season. Spreading of dark dust of various kinds is done by plane or on flat areas by spreaders towed behind tracked vehicles.

Canadian experiments have shown that the most suitable particle size for ice melting is from .04 to .10 inches (1 to 2.5 mm.) in diameter. It is best applied with a density of between .04 and .08 pounds per square foot (200–400 grams per square meter).

If new snow falls on top of the dust the melting effect may cease, but sometimes solar radiation penetrates the snowcover and the dark particles continue their job of melting though buried. Melting can go on under cloudy skies also. As long as radiation gets through to the particles, even in reduced amounts, melting can occur.

Sand, it was found, left a pitted, uneven surface with melt holes as large as 16 inches deep and 4 inches across, while fuel oil spread on the ice caused increased melting at a uniform rate which left a smooth surface. Mud is sometimes pumped up from the bottom of a frozen lake and spread in a thin layer on top of the lake ice.

There is considerable interest in this area of experimentation in the United States too. Way back in 1761 Benjamin Franklin was conducting snow-melting tests. He placed squares of different-colored cloths on snow in the sun, and in a short time the darker squares, especially the black one, sank lower in the snow than did the light-colored squares. The white one sank least of all. He used this information to advise a woman that summer hats should be white because that color repels heat best.

Coal dust has been sprinkled on the soil at Fairbanks, Alaska, in the spring. This increases soil temperatures to a depth of about four inches and gives plants better root systems. Experimenters have suggested the use of planes to spread the dust at a recommended 1,000 pounds per acre.

Melting of snow with carbon powder was tried at Donner Pass in California's Sierra Nevada at 7,000 feet. It caused the sun to melt snow 1.6 times as fast as untreated snow in May. And at 12,000 feet in July carbon powder caused snow to melt 2.4 times as fast as untreated snow. Ground-up tires were also used in the tests.

In the United States all glaciers east of the Rockies, and those of Hawaii, disappeared long ago. The glaciers of the Rockies—in Montana, Wyoming, and Colorado—are relatively small and most are mere hangers-on. So are many of those in the mountains of the coast states—Washington, Oregon, and California; there glaciers are larger and more numerous toward the north. Between those mountains and the Rockies there are few glaciers to be found. The southernmost glacier in the United States is a tiny cirque one in the southern Sierras of California. It is kept alive only because snow falls into its catchment basin from slopes above. It is not the southernmost glacier in all of North America, notwithstanding the California Chamber of Commerce. Glaciers still hold on to Mount Popocatepetl and Mount Ixtacihuatl south of Mexico City at latitude 19° N.

The largest glaciers, and 90 per cent of the glacial ice, remaining in the United States outside Alaska are in the state of Washington—about 135 square miles of ice—where humid Pacific air rising over the Olympic and Cascade mountains maintains them. And in that state glacier-melting tests have been conducted.

Washington's Division of Power Resources has run artificial-melting experiments on an unnamed ninety-acre glacier in the North Cascades "to better understand hydro-meteorological relationships and to promote water resource development techniques," or, stated another way, to learn how to increase the level of guaranteed hydroelectric power by extracting extra water from glaciers when streamflow is deficient. Test sections were only small patches. They were dusted with coal dust and carbon black, and melting was measured by weighing the runoff. On warm, sunny days melting was increased from 12 to 144 per cent. A federal power agency has studied the glacier-melting tests with a possible practical application in mind. Glaciers would be melted, it is said, only when streamflow was abnormally low, or when it was necessary to fill a large reservoir quickly. Would that be one year in five? Or, as population increases, every other year? Or every June? Or June, July, and August? Only in emergencies perhaps. Would a falling off of profits be a legitimate emergency? A power shortage might.

Glaciers in Oregon, California, Nevada, Colorado, Wyoming, and Montana, it is said, may in the future be "controlled" by methods that Washington has been testing. Once glacier melting is permitted, it will be difficult to curtail. Where it is practiced the extra water will have kept consumer costs down and attracted more families and industries. They will have become dependent upon the glacier melt-water. Even if a glacier were seen to be disappearing, the citizens could not be left high and dry. Who could stop the established practice of melting glaciers? Once the precedent was accepted, once the "ice was broken," it is more likely there would be pressure to melt additional glaciers, if they still remained.

We hear nothing about cleaning glaciers after dusting experiments. That would be a nearly impossible task. It might be argued that cleaning them is unnecessary because the next heavy snowfall would cover and inactivate the dark material. But the next thaw, or some later thaw, would expose it again. It could remain there capable of doing its destructive work for an indefinite period. It could become active again at an unpredictable time when melting was faster than desired already. One can imagine cases of "controlled" dusting where a glacier is dusted and then snow covers the dust, so the glacier is redusted and it snows again, and so on. Glaciers could receive any number of dust layers. The job of undusting glaciers would be vastly more difficult than flying over in a plane and dumping dirt—and probably not even possible—but the full consequences of dusting should be thought about. If those who dirtied glaciers had to clean them, if those who took away part of a glacier had to restore it, glacier melting would not seem as lucrative nor as attractive a solution to water problems as it does now. Sometime soon we must stop letting privileged groups do what they wish to our landscape and the geographical balance and then walk away with full purses, bearing no responsibility for the shambles they leave behind.

No one can tell what climatic conditions will prevail in the future. Natural melting of glaciers may increase. Imagine what could happen if years after glacier dusting was done the buried dust became exposed, layer after layer, and melting got out of control. There would be no way to extract the dust or slow the melting then, for the dust would be irretrievably mixed with the ice. Facilities downstream might not be able to handle the increased volume of water. The floods could be horrendous and could continue unstoppable for years.

Those who want to melt glaciers—in the United States, at least—say they do not want to destroy glaciers, and this is so, for to do so would hurt their enterprise in the long run.

Melters who would rob a glacier trust that nature will replenish the loss, and if it does not, then they say they will try to bring the glacier back to its former size, using snow-catching and snow-preserving methods and cloud seeding. With cloud seeding they would like to be able to force the atmosphere to drop snow and to put that snow back on the very glaciers they have depleted. What no one can honestly claim to have done even once—with luck—they propose to do repeatedly on a regular, controlled basis. Maybe sometime in the future such a promise might be kept, but not now.

Although *Ice Age Lost* places emphasis on the prospect of the deterioration of glaciers and the disappearance of remaining Ice Age environments and biota, to be objective and realistic it should keep readers aware of the opposite prospect, of the still-existing strength of glaciers and the Cold, and the possibility of the return of Ice Age conditions. This alternate viewpoint is looked at here briefly in the next chapters—so that readers will not be blindly restricted to the point of view stressed in this book; so that they may receive a full, multi-sided survey of Ice Age and post-Ice Age conditions; and so that they can appreciate and capture the feel of glaciers and the Ice Age in every possible way. After side-stepping to explore this alternate viewpoint we shall return to the primary "lost" theme.

CHAPTER 13

INVITING GLACIERS BACK

In only a moment we both will be old.
We won't even notice the world turning cold.*
From the song "MY CUP RUNNETH OVER"

THE writing of this book began one day in late November. Experiencing a common frustration of writers, I sat at my typewriter and looked out my apartment windows, waiting for those first elusive words to find their way onto paper. The Indian-summer morning was bright and blue, and the sun, though low, was drawing the chill out of the air. My sentences were unacceptable, my ideas wayward. By noon the clear air had warmed and I was musing whether this was a year when winter would come late. (It usually strikes in earnest around Thanksgivingtime here in Wisconsin.) The almost balmy breeze that drifted through my casement windows was not stimulating. I gave up trying to write for a while. By mid-afternoon the sky was graying with thickening sheets of clouds. A different, stronger wind began tearing with determination at the corner of the building. People down on the street, dressed too lightly, crossed their arms tightly and hurried along.

One never knows just when the balance of power will shift, when dallying Fall will have its bluff called and will bow to the flailing wind-whip of Winter. Fall had fought a gallant holding action, but Winter, doggedly on schedule, had chosen this hour to make its triumphant entrance.

There was new energy in the air. Words began stringing themselves

* Copyright 1966 by Tom Jones and Harvey Schmidt. Used by permission of Chappell & Co., Inc.

out on the paper now in paragraphs. It was thrilling to feel Cold assert its old power. How much more thrilling to think of those mightier winters of the Ice Age past when the greatest cold of all descended upon the world.

Another glacial stage of the Ice Age could come upon us as easily as winter comes, just taking longer to effect its change. Or we could as readily slip further into a post-glacial epoch of continued warmth. Whether this intermediate, medium-warm, vacillating period we live in is a "spring" or a "fall" only time will tell. If it is a spring then it is leading up to a summery, more tropical future, either to a milder-than-now, true interglacial period or to a lasting post-glacial period which could say good-by to glaciers for good. Or if this intermediate period is an autumn, then there will come a shift to the cold end of the climatic teeter-totter, and the ice could again begin to have its way.

If we are destined for another glacial stage the push that sends us there might be one over which we have utterly no control, or it might be one that people themselves initiate. The latter is by far the less likely.

If glaciation was caused by some happening in outer space—like a reduction in the sun's radiation, or Earth's passing through a space cloud which lessened incoming solar energy—and if that situation should recur, there is nothing we could do to forestall it.

Even if glaciation was caused by conditions right here on this planet, conditions that were likewise unalterable—like increased volcanic explosions which darken the sky, or shifting of Earth's axis, or drifting of continents—then its coming would be equally unstoppable at this stage of our capabilities.

But among the many geographical conditions that affect climate there are some we do have the ability to modify, knowingly or unknowingly. During the many warm-and-cold fluctuations of the Ice Age prehistoric people's influence on environmental controls was as negligible as horses' or mammoths', but people of today, because of their far greater numbers and their mechanical and chemical know-how, have it in their power to alter Earth's geographical make-up, and in doing so to modify local daily weather in some ways, and perhaps long-range climate as well. The human animal now also has motivation—desire for profit, for improved living standards, and for elbow room—to change large tracts of a continent. Lesser, localized changes in our habitat register a cumulative effect when there are enough of them.

Consider that forests help retain moisture, slow surface winds by friction, retard the heating and cooling of the ground, shade and protect snow that falls in them, and transpire tremendous quantities of moisture to the air. And consider how dark surfaces absorb more heat than light ones, and how barren surfaces will heat rapidly and send up strong convectional air currents which can lead to cloud and precipitation, and how they in turn cool more rapidly than surfaces insulated by vegetation. And bear in mind how water bodies humidify the air, and moderate temperatures because of their exceptional ability to absorb and store heat, and how they at times produce fog which further lessens the likelihood of extreme temperatures.

Then consider how weather is modified by clearing away forests over wide areas, or replanting them; by plowing up green plains across a whole country and making them dark-brown croplands; by covering natural earth with strips and plots of bare concrete; by erecting clusters of buildings to the dimensions of high hills; by creating lakes where none were, or by draining lakes and swamps out of existence.

It is not known whether climatically an ice age is turned on initially by a temperature drop or by increased moisture availability, or more vigorous atmospheric circulation, or a combination of these or other factors, so there is no certainty that changes in Earth's geography that seem wholly divorced from the fate of glaciers do not indirectly bear upon it. Certain changes that are not by themselves instrumental in inviting another ice age may disturb a climatic cycle just enough to give glaciers a slight advantage in critical areas and allow them to start their comeback.

Later we shall look more closely at our increasing ability and hankering to modify weather and climate, but here it is just pointed out that this ability, though still limited, is real.

Besides redesigning Earth's surface we are fouling the air. Atmospheric scientists warn that by contaminating the atmosphere we may be altering our climate, but they cannot tell whether we might be warming or cooling it. By adding carbon dioxide to the atmosphere, which increases the greenhouse effect, we tend to raise temperatures; but, on the other hand, the tiny solid particles of dust and smoke which are being expelled and churned into the air as a result of human activities create a haze that could be screening out a fraction of the sun's energy and so be lowering the planet's temperature by making its atmosphere less of an open window and more of a curtain. These two opposing effects—one warming and the other cooling—could be canceling each other out, or one could be having a greater effect than the other, or both could be inconsequential. We just don't know yet, for we cannot measure the effects quantitatively.

The National Center for Atmospheric Research in Boulder, Colorado, has stated that dirty air is apparently causing the temperature of Earth's air envelope to drop, that the average temperature seems to have fallen one half degree Fahrenheit in a recent seventeen-year period, according to available data. It is said that a temperature drop like that should move the frost line 100 miles equatorward. Other studies of world weather statistics also have shown a small average temperature decline in recent decades. But how truly representative are the statistics?

Determinations of climatic trends over wide areas that are based on recorded weather data alone should be looked at with caution and reservation. Weather records around the world are too patchy and meager, and have been kept for too short a time to show an overall climatic trend that can be considered valid. Therefore, the extent to which humans may have been modifying world climate artificially cannot be clearly ascertained either.

Long before the Industrial Revolution began polluting the air in the late eighteenth century, the effects of dirt-darkened skies were experienced, though intermittently, with clouds from volcano eruptions and dust storms. The latter may have been partially the result of Man's disturbing the natural habitat.

Dust storms do not seem to alter established worldwide climate, although continual dust can affect local climates significantly. During the warming latter part of the Ice Age there were greater and longer-lasting dust storms than now, judging from the deep and widespread accumulations of loess that were wind-deposited then in North America, Europe, and China—and still the temperature rose; but here we shall not try to compare those dust storms, about which knowledge is limited, with those of the present. Perhaps if dust palls increased considerably in various parts of the world they would become a factor to be reckoned with; and Man is capable of making them increase by continuing careless encroachment upon dry lands around the world.

Overgrazing and overcultivation of dry lands have denuded large dust-bowl areas from which storms blow clouds of light soil particles thousands of miles to regions far afield. Some land that is barren desert today may have become so not just because of climate but because animals in too great numbers tramped, grazed, and browsed upon the land which had only a precariously thin vegetation cover. Once gone, the vegetation could not re-establish itself and the land lay victim to the wind.

Although dust storms are not thought of seriously at this time as a possible cause of worldwide climatic change, civilization's air pollution is. Airborne refuse has been accumulating in ever-increasing amounts, and much of the polluting material drifts to the atmosphere's upper levels where it is not dissipated but makes a filter of the whole spherical sky.

Metropolitan areas were the first to feel the convincing effect of pollution as gray tents of smog camped over them. Research on cloud coverage over ten large American and European cities shows a 50 to 60 per cent increase in gray, sunless days during the past century.

Some scientists believe that modern civilization's exhalations which float between us and the sun could bring the temperature drop necessary to usher back the Ice Age. Judging how many glaciers are advancing and how many more are at or near equilibrium, that drop would not have to be great. At any rate, any slight drop would bring us closer to the brink.

It may be that the determination of whether or not glaciers will grow again—or diminish—is as much our doing as is the remodeling of the landscape and the transparency of the atmosphere. If so, then we—earth scientists, engineers, teachers, trainers of children, legislators, farmers, voices of public opinion, the voting public—have some say-so over the destiny of glaciers.

Each glacier is an heirloom which Earth wears like a lustrous, precious pearl. Even more precious than pearl is a glacier, for it is not only beautiful to the eye and thrilling to the touch and aglow with the wonder of its formation, but besides being all that it is made of the most valuable and vital of substances—pure, life-giving water.

Enterprising pearl dealers learned to stimulate pearl growth artificially by hand-planting bits of irritating material inside oysters' shells and then waiting several years for the oysters to deposit smooth layers of pearl around the irritants. And enterprising farmers have learned to culture glaciers too by planting pieces of ice in the mountains and helping them add layers to themselves and develop into lasting glaciers.

Farmers of Pakistan long ago worked out an elaborate technique for growing glaciers in order to keep their streams flowing during dry spells. A glacier is a water reservoir of the cheapest kind. Rain runs off the land as it falls. Snow stays longer but often not long enough to help pastures and crops survive a hot summer. Streams that are fed only by rain and snow-melt have low volume some years and often dry up before the end of summer, but those that are glacier-fed are more dependable, having a slow-melting reserve that delivers water even during drier-than-usual spells. A glacier is money in the bank, a trust fund that pays reliably, steadily year after year. Building a glacier is not easy. The patient, tedious work these people do in the hope of starting one demonstrates the value of glaciers. They are wealth. They are life.

The art of glacier propagation seems to be best developed in arid mountain regions of northern Pakistan, particularly the Karakoram and Hindu Kush ranges, which supply the irrigation water on which all local farmers and villagers depend. From separate reports of Elizabeth Staley and Patrick Fagan in *Ice* a picture of the glacier-making operations can be put together.

When the glacier that supplies a village or area disappears, or when an area that has been relying only on snow melt wants its own glacier, the inhabitants cooperate in making one. The process is not always successful, but apparently it succeeds often enough to keep the practice going. This is a borderline area for glaciers. If local conditions are just right a glacier can survive. It would be hard for a glacier to form naturally here, but if given a start and protected for a while it can sometimes take hold. As we have noted before, a glacier tends to create its own environment, so once it exists it helps sustain itself and may grow.

This is the traditional recipe in case you ever own your own mountain and would like to whip up a little glacier.

To begin, take some male glacier ice and some female glacier ice, bed them down together, and if all goes well in time you will have yourself a little glacier. If you are dubious you could ask the Pakistan villagers, and they would explain. Male glaciers are clear ice, they say. Female glaciers are cloudy ice and they are wetter. You take some ice from each in the proper proportions—two parts of female ice for one part of male ice—mix the ices together, and plant them in a cool, shady pit as high as possible at a tributary head of an irrigation stream, around snowline elevation. The pit is lined with such substances as snow, straw, herbs, dung, salt, or charcoal, and the ice is covered with this too. There may be several alternating layers of ice and insulating material. About eighty pounds of ice is buried to start with. Cold weather or heavy snows help the glacier to form, but if weather does not assist, more ice is added from time to time.

Coolies carry chunks of ice to the spot, being sure to use ice of both sexes even though one type can be obtained nearby and the other has to be brought from a great distance away. One report tells that on top of each ice layer leather bottles containing a certain kind of water are laid along with thorns which will prick the bottles now and then to let water seep out. On top of the uppermost ice layer numerous earthenware jars are set, containing the same special water. They break when the water freezes and release water gradually so it can adhere to the ice.

This operation of planting ice may continue several years, and if the pit area can stay cold enough to preserve the ice that is there and slow down the melting of snow that falls, a glacier may get started.

There are more sophisticated ways of propagating glaciers. Glaciers really can be made, but not everywhere. The environment has to be right to begin with, and then they have to be nurtured and protected in order to become established, and even then, being in a marginal location, they are in danger of failing at any time, unless climatic conditions shift definitely in their favor.

Japan's Nagoya University started a glacier-making project a few years ago. Although Japan has lost all its glaciers it still has heavy snows, and firn (the tiny ice grains formed from old snow) exists in high mountain areas of central and northern Japan. It indicates that the snow can resist melting for several years. A basin in the Tateyama Mountains, about 36° N, containing firn over fifteen years old has been used for the university's experiment in artificially producing a valley glacier. The researchers hope to study progressive changes from firn to glacier ice and to see whether the valley glacier might serve as a "water reservoir."

The plan is to increase winter snow accumulation in the firn area. Drift fences are erected in such a fashion as to make the snow settle in places where it will best help the glacier grow. The ridge above the firn area is wide enough to allow bulldozers to push snow down from it onto the firn field below. Explosives are used to make snow avalanche from the sides into the firn basin. Melting and evaporation of the amassed snow and firn is retarded by a covering of plastic sheets.

Foresters and watershed managers in many countries utilize similar and additional methods to conserve snow in the highlands and make it collect in locations where it will last longest.

Water-supply engineers have given thought to creating artificial glaciers by pumping water into shaded mountain valleys in winter. If glacier makers prove to be successful one can foresee communities hiring them to, say, establish a glacier on heights overlooking their town or resort as a tourist attraction or local beautification project. Or they could have glaciers built to provide a more dependable water supply. That does not sound any more impractical or impossible than constructing enormous dams, or creating lakes to attract vacationers and land buyers, or managing multi-state watersheds, or building brand-new towns in the middle of nowhere.

As the benefits of glaciers come to be recognized how many more will be propagated? Will whole glacier farms be established with each glacier helping to preserve the others? Will small glaciers be given growth stimulants? Could glacier growth get out of hand?

It should be easier for glaciers around the world to stage a comeback than it was for them to get started in the first place, either at the beginning of the Ice Age or at the start of any of the subsequent glacial stages following long interglacials, simply because the ice now has a strong foothold. Most high mountains around the world have glaciers on them; or firn, ready to make glaciers if given a few thawless years; or other stationary ice. And most of the other high mountains have snow part or all of the year.

All in all, there is enough snow and ice around to give an ice age a good start. In existing glaciers there is enough ice to form a layer 400 feet thick over the whole land surface of the earth. There is much ice left from old glaciers which we cannot even see. Moraines and outwash and terraces around many lifeless glaciers contain large relict bodies of ice. One moraine in Norway contains ice that is 2,600 years old. Soil has developed over many hibernating glaciers and sometimes vegetation has taken root too, helping camouflage them, covering them so they appear to be solid ground. The huge, lazy Malaspina on the Gulf of Alaska, a piedmont, tidewater glacier larger than Rhode Island, has forest growing on it.

In accessible places where Ice Age glaciers used to be but are no longer, snowdrifts are watched with more than casual interest. They could be omens of glacier rebirth. If snowdrifts should last over summer very many years in a row, a glacier might be in the making. For a long time British glaciologists kept year-to-year snow surveys to see where snow lingered longest in the British Isles. World War II interrupted those surveys, but right after the war they were resumed, and a reminder appeared in the *Journal of Glaciology* to that effect:

It is particularly hoped that observers may be found who can keep a watchful eye during the summer months on the quasi-permanent snow-beds of the Allt-a-Mhuillin (Ben Nevis) and Braeriach corries, and on the long-surviving drift in the deep gulley beneath the summit of Carnedd Llewelyn. A question to be investigated is whether or not there remain to be discovered other northern hollows that usually or often harbour snow from year's end to year's end.

Scotland's Ben Nevis is the highest peak in the British Isles—4,406 feet; and its Gaelic name, now shortened, means "the mountain with the cold brow." It is noted on maps as a place where snowdrifts of some size have persisted through the summer in a majority of recent years. From Ben Nevis and nearby summits flowed glaciers that helped make the ice cap that covered northern Great Britain. Its glaciers may have been the first to form on the island, so others besides poets, painters, and tourists study famous old Ben Nevis with a contemplative eye.

The place where snow is reputed to last longest in Great Britain south of Scotland is the gully, y Ffos Ddyfn, near the top of 3,484-foot Carnedd Llewelyn. This is strange because that gully faces almost due south.

Australians are carefully watching for any trend toward glacier re-formation in their Alps. A semi-permanent snowbank on Mount Twynam has been under regular surveillance since 1962. On the tropical island of Hawaii fossil ice which may be left from the Ice Age has been found atop volcanoes.

In northeastern United States on New Hampshire's 6,288-foot

Mount Washington snow and ice manage to last through the summer, not in continuous surface cover but in niches among boulders and in cave-like recesses and ravines. Skiing is possible well into June. Permafrost exists in the ground in places. Mount Washington Observatory informs me, "We are only a degree or so away in our temperature normals from a return of glacial conditions."

Climatologists and glaciologists watch the rising and falling of snowlines on mountains for indications of climatic change, and check glaciers wherever they can, within physical and financial limitations. Glaciers are marked with stakes to show their movement; the melted runoff is measured; they are surveyed and photographed. Their height as well as length is noted periodically, as is the amount of new snow and firn, in order to tell whether they are becoming healthier or weaker. While the consensus holds that glaciers around the world are generally retreating, a great number are advancing too, and we do not have sufficient data to know what the average glacier fluctuation amounts to worldwide.

Antarctica, which holds most of the world's ice, is the biggest question mark. Crews of scientists there are gathering data on snowfall and glacier ablation to learn what the budget of the White Continent is at this time—whether it is receiving enough snow to maintain or enlarge its present glaciers, or whether it is losing ice faster than it can manufacture it. Some glaciologists, you will recall, think the expansion of ice on Antarctica is what steers the world back on the track to glacial conditions.

In some places where glaciers disappeared completely at the end of the Ice Age new glaciers have formed in recent centuries. Also, there are many marginal sites where glaciers cannot quite sprout spontaneously but if the seeds were dropped and nurtured awhile glaciers could take root.

Atmospheric scientists have been experimenting with seeding clouds in certain mountain areas to produce snow because when the mountains' snow supply melts away by midsummer, rivers and irrigation channels may be dry the latter part of the growing season. If snow could be packed heavily enough on the mountains to last through the summer this problem would be alleviated—and glaciers might begin. If they did, they would probably be heartily welcomed by those who want steady streamflow.

Another reason for wanting to increase snowfall in mountains is to justify artificial melting of snow and ice. Advocates of snow and ice melting argue, "What we melt we shall have the snowmakers put back by seeding clouds and artificially inducing snow." But if they should be successful in forcing snow could they control the amount that would fall? Clouds might release more snow than was called for. And could they make the snow fall where it was requested and not somewhere else? Sometimes orders for snow might have to be re-delivered if the first snow deliveries went astray, and there could be quite a lot of free snow piling up where it was not wanted without anyone really knowing whether it was natural or artificially caused. As the need for more and more water grows one can foresee orders—even "advance orders"—to deliver more and more snow. Glaciers could begin where glaciers are not now, and little ones could become bigger ones in a short time.

Cloud seeders are not the only ones in the snowmaking business. While success of their method is still dubious, there is another method used right on the ground whose results are apparent. The reference is to snowmaking machines. The amount of snow they manufacture is infinitesimal on the world scene but it deserves mention. Their operation has seen such phenomenal success that large-scale applications of this process may develop. They have been a boon to ski resorts around the world and are used not only in areas where snow is scanty or thaws common, but also in areas of heavy snow. Even in Alaska! No one is asked to believe that snowmaking machines will generate an ice age nor at this point even a glacier, but added snow does affect the micro-climate of a locality. It seems in-evitable that these machines will be used for larger jobs than coating ski slopes—perhaps paving snowmobile courses, creating "winter wonderlands," or in "agribusiness" providing beneficial snowcover for acres of cropland.

Many mountain playgrounds that our society calls its own really belong to the snow, for it withdraws only briefly in summer. Many inns in parks and resorts, which in summer hold throngs of tourists and are alive with lights and music and sports, must rush to close in early September and do not reopen till June. In winter they are smothered in snow up to the roofs.

If glaciers should for any reason start re-forming in regions where the great ice sheets had their inception, they might have assists that earlier ones starting out after true interglacials did not have—better moisture sources in key areas. Northeastern North America, where the great Laurentide Sheet formed, now has more water surface to draw moisture from. Hudson Bay exists because the land is still depressed from having been weighted down by the last ice sheet of the Wisconsin Stage. Probably some of the water around the Arctic islands also used to be dry land. When the bay and these waters are unfrozen they provide moisture to land about them. Hudson Bay's southernmost part, James Bay, strikes deep into the continent's interior as far as 51° S. Jump only about two degrees farther south, about 150 miles, and there is the north shore of Lake Superior, largest fresh-water lake in the world. Together the five Great Lakes constitute the world's largest group of lakes. They were not there either when the old ice sheets were forming.

Of course, the main moisture supply would come from the oceans, but the added increment from the Great Lakes would not hurt and might facilitate the ice's southward growth by making snowcover in that mid-continental area heavier than it otherwise would be. The Great Lakes are good snowmakers. Dry winter winds blowing out of Canada evaporate moisture in crossing the lakes and release it easily when lifted. Lake Superior's southern shore is a snowbelt. Northern Michigan, surrounded by lakes, receives more snow than any place between the Rockies and New England. A winter's snowfall on Michigan's Keweenaw Peninsula often amounts to over 250 inches. Snow squalls are frequent along Lake Erie's southern shore. The Buffalo area of western New York between lakes Erie and Ontario takes the brunt of winds and snowstorms coming off the water and expects heavier snows and more traffic tie-ups than places farther from the lakes.

In northern Europe the Gulf of Bothnia, the Baltic Sea's northward extension, occupies land left depressed by the last ice sheet. There the ocean now extends up the center of Scandinavia, where the ice sheets began, so there too the moisture supply might be greater than when the earlier glaciers formed. (In about 2,500 years dry land will appear in the rising Gulf of Bothnia and later the gulf will shrink to just a small lake on higher land.) Add to this situation the fact that some Swedish meteorologists have been saying that snowfall over the mountains of central Sweden could be substantially increased by seeding super-cooled mountain clouds in winter. Extensive clouds over Scandinavia often give no precipitation, and there are proposals to seed them "to provoke snow on the mountains."

Glaciers may at any time get the go signal from whatever quarter it comes. If key areas are given extra moisture assists glaciers might get off to a better than usual start. The ice cover on the Arctic Ocean has lessened in recent years too, exposing more open water. The highlands where the ice sheets last formed are still there and some have continued rising even to the present, those of western North America, parts of Asia, and the Andes among them. If glaciers should be invited to reappear there may be nothing holding them back.

If it should happen, if mountains should start housing new glaciers in old cirques, if snowlines fell lower around highlands, if all the amorphous, ghostly shapes of still glaciers stirred in their white shrouds, came back to life and moved toward us from all directions, would we do anything about it? Could all the powers of the modern age stop the ice? Could glaciers be stopped by physically attacking and destroying them? Surely thermonuclear bombs could shatter and turn whole glaciers to steam—or could they? No one has ever tried, so no one knows what would happen, and the Atomic Energy Commission says that at present no research efforts are being applied to the use of nuclear explosives in ice.

Initial encroachments by glaciers would not cause alarm because glaciers fluctuate erratically all the time. Even if a general advance were evident nothing would be done except where a particular glacier bothered some community. No general war on glaciers would be declared. A grievance usually has to be voiced by a great number of people, loud and long, before concerted action is taken if it requires releasing money, thinking up a plan, upsetting the routine, and expending physical energy. By then things might be glacially well under way.

Out of curiosity I made an inquiry about the possibility of combating an actual glacier threat. I asked the Department of Highways of Alaska what it would do if a glacier encroached upon an important highway, for such a situation has come close to happening. The reply from the Maintenance Division was: "If a glacier were to advance across a highway, we would undoubtedly be forced to leave the existing route and rebuild elsewhere. The force exerted by a glacier is quite impressive when seen first-hand, and certainly beyond our financial resources to stop."

It is not easy to undo glaciers or keep them in bounds. They can endure in the unfriendliest of places. Like on volcanoes, for example, where they withstand blasting and heat. In Iceland sometimes catastrophic floods occur when glacier-covered volcanoes erupt but the glaciers persist. Glaciers capped Mount Katmai in southwestern Alaska when in 1912 that mountain volcano erupted and blew its top off, glaciers and all, but the glaciers re-formed *inside* the 2,600-foot deep crater of the topless peak.

After the 1964 Good Friday Earthquake shook south-central Alaska there was intense curiosity to see what damage it had done to glaciers, for the region where it occurred contained many hundreds of glaciers and the area hardest hit was about 20 per cent ice-covered. And this was a tremendous quake, the strongest ever known to hit North America, and the second most powerful ever recorded. (A worse one hit Chile in 1906.) Its energy was 200,000 megatons, more than 2,000 times the power of the mightiest nuclear bomb yet exploded and 400 times the total of all nuclear bombs exploded to that time. The earth was rent with cracks and fissures as 24,000 square miles of this mountainous area were raised or lowered an average of three to eight feet. Whole towns crumbled or sank. The main street in Anchorage dropped twenty feet. River waters were diverted. Trees were whipped back and forth so violently their tops hit the ground on either side. Harbors were beaten to ruins by tidal waves. Thousands of miles away needles jumped off seismographs. Two thousand miles away in California a nine-foot tidal wave smashed a city's business center. Water sloshed out of swimming pools in Texas. Rock avalanches rattled down from the Alaska mountains that felt the destructive power of this record-breaking earthquake. And what happened to the glaciers held in those mountains?

Glaciologists went to see, and the word they brought back was astonishing. The glaciers were practically unscathed!

William O. Field of the World Data Center for Glaciology of the American Geographical Society inspected them and said they were "extremely stable." Still, a wait-and-see attitude was taken. Maybe aftereffects of the quake would appear. Study of the earthquaked glaciers continued. The following year reports of careful study began to appear.

R. H. Ragle, J. E. Sater, and W. O. Field, who made reconnaissance flights, summarized their findings for the Arctic Institute of North America. Their report cites landslides, avalanches, and severely cracked lake ice as evidence of the quake's strength, but includes these striking comments about the glaciers:

The scarcity of obvious change was surprising since the glaciers must have been shaken violently by the earthquake. . . . With few exceptions, hanging glaciers did not appear to have been affected and there was no unusual calving of glacier termini into tidewater.

Austin Post made a comprehensive study of the quake-rocked glaciers for the U. S. Geological Survey, and in his well-studied conclusion, written almost a year and a half after the event, he states:

No clearly defined large-scale dynamic response to avalanche loading or other effects of earthquake shaking had been found in any glacier. . . . No apparent changes in ice-dammed lakes or glacier drainage resulted from the earthquake. Tidal glaciers show little immediate effects.

In the *Journal of Glaciology*, L. E. Nielsen wrote that while some steep ice on slopes looked unusually shattered after the quake, no ice avalanches were found that were large enough to appreciably affect any glacier's regime.

That observation is similar to one made after the extremely strong earthquake of 1899 near Yakutat Bay, Alaska. Post reviewed that event:

It is pertinent to note that the three steep hanging glaciers on the mountain west of Disenchantment Bay were not shaken down by the 1899 earthquake although this area was severely shaken and the coastline raised 13 meters. . . . This occurrence provides additional evidence that even very steeply pitched glaciers are not greatly affected by violent earthquakes.

Glaciers are not things you can simply get rid of when you do not want them. If we invite them back we cannot just shoo them away again. If they did take to growing, even on their own, it is only human egotism that says they could be stopped. With our nuclear bombs, the most powerful destructive weapon we have, we may think we could bombard glaciers until they are gone, but consider how many glaciers there are now, and how many more there could be by the time such extreme measures were called upon. Man-made power is puny against nature's. The energy of a Nagasaki-type A bomb is expended in a thunderstorm every ten seconds. To attack all glaciers would be to carry on a major world war continuously for years on end, a war more extensive and longer than any ever waged. An endless war against glaciers would not be financially feasible; and there could be atmospheric effects from bombing, including fallout, that would make such operations impossible over a prolonged period. Who knows but what the towering, bulging clouds resulting from the blasts, with the evaporated moisture that went aloft, might greatly increase snowfall in various places, refeeding glaciers and starting new ones.

We always think that a few years hence we shall have much more capability and be much further advanced scientifically so that what we cannot do today we can do tomorrow, however impossible it seems. It never occurs to us that we might be at our zenith now, or near it. Our superman civilization (which is generated from just a few elite parts of the globe) could slow to a standstill or decline from many causes—economic collapse, epidemics, laziness, lawlessness, or a warring away of vigor and wealth. All the while glaciers would stoically move on like fearless robots whose supply lines were uncuttable, who healed when wounded, who if removed could reappear out of thin air. The endless hordes of Asia would seem toy soldiers beside them. The words of anthropologist-philosopher Loren Eiseley resound ominously: "Man's survival record, for all his brains, is not impressive against the cunning patience of the unexpired Great Winter."

If glaciers were returning there is no Great Wall, no Maginot Line, no Dew Line that could hold them in check. Farmers in the bouldery hills of New England say, "The Ice Age isn't over. The glaciers have just gone back for more rocks." Perhaps they have. Next time they set out to conquer the world—if there is a next time—we like to envision them meeting a powerful foe that was not present during their previous campaigns: a people armed with intelligence, sophisticated weaponry, and foreknowledge of their often-used battle plans and invasion routes. An equally likely prospect is that if and when the glaciers come again—sometime far in the future—they will find the world just as wild and primitive and unaware as it was each previous time they came!

But here as we consider just the immediate future, we see Man stepping up his experimentation with glaciers. In the past they were controlled by nature alone, and like trained performing animals they have been going through their paces as directed by that one master only. Now other masters prepare to step into the arena and give orders and upset the routine. We step across a critical line when we begin to tamper with the equilibrium of glaciers. We know how to make them shrink and are learning to make them grow. We cannot control their responses but we are cracking the whip. And just what do we expect to happen?

No two glaciers act the same. Twins side by side can behave as differently and as unpredictably as any two brothers or sisters. What we are about to do amounts to nudging a sleeping beast just to see what it will do, or using a signal whose response we cannot anticipate, or pulling a string without knowing what it will unravel.

The fact is, we are extremely ignorant about glaciers. We don't know how they came to be here. We don't know how fast their arrival occurred nor how fast they could disappear. We are upsetting their balance but we don't know what the effect will be, on them or us. Often their true response is delayed for years. The one thing we know for certain about glaciers is that they do react—often quite dramatically.

CHAPTER 14

THOSE GALLOPING GLACIERS

Some say the world will end in fire,
Some say in ice.*

ROBERT FROST

TODAY'S army of glaciers is composed mainly of slow plodders trudging like weary, footsore troops. If their positions are noted once a year or every few years that is usually frequent enough to determine a change of position, if any. Many lethargic glaciers show little change from one decade to the next. Each glacier moves at its own rate regardless of what neighboring ones do. But every once in a while out of the ranks a glacier will unexplainedly blitz ahead—relatively speaking—as though it had gone berserk. Its pace will not be so fast as to kick up a cloud of dust. It is still stealthy. Like a shadow creeping over a lawn, it is too slow to notice in one glance but apparent after a series of looks. And as such a glacier takes off like a hare among tortoises, melting cannot check it as fast as it enlarges.

There is just no telling when a glacier is about to surge ahead like that. Sometimes one looks as though it is ready to and then it does not, while some inert-looking one will. Some have a reputation for surging occasionally, but with no predictable periodicity. The surging of others comes as a complete surprise.

Glaciers take on a whole new appearance when they surge. Ice streams that were smooth and tranquil break up in turmoil. A wave of ice bulges to the front and the glacier's surface buckles chaotically

---

* From "Fire and Ice" from *The Poetry of Robert Frost* edited by Edward Connery Lathem. Copyright 1923, © 1969 by Holt, Rinehart and Winston, Inc. Copyright 1951 by Robert Frost. Reprinted by permission of Holt, Rinehart and Winston, Inc., Jonathan Cape Ltd., and the Estate of Robert Frost.

into peaks and spires and deep cracks. In valley glaciers the straight
medial moraines, those dark lengthwise streaks, swerve in slow motion in contorted curves as they are shoved aside by the irregular ice
flow. Normally, medial moraines are straight or gently curved like
lane lines on a highway, separating parallel tributary glaciers that
have merged and are flowing alongside one another. But when one
tributary of ice surges against another there is a pile-up and the medial moraine is pushed aside in bulges and tight bends.

Surging of glaciers is nothing new. This spontaneous iceflooding
must have been going on ever since glaciers came into being. But
only in recent years have we come to realize how frequent an occurrence it is, and only of late has this phenomenon been given high-
priority attention in glaciological circles. Earlier only people living
near glaciers saw the surges, and whether frightened or unimpressed
they accepted such events as being as natural as an avalanche or flood.
The news was not flashed abroad. Occasionally in the case of a
really disastrous surge the story reached the outside, but still little
more than regional attention was given to it. Or it might rate a mention in someone's travelogue.

Over a period of time those whose ears were tuned to glacier
idiosyncrasies came to realize that this surging was a significant trait
of glaciers, and, in fact, one of their most interesting ones. Out-of-
sight glaciers which are thought to be doing nothing unusual for years
may be having occasional escapades after all.

Reports of surging glaciers are as teasing as rumors of scandal.
They leave one wanting proof. Tales would be heard of a glacier that
had staged a sudden "catastrophic advance" (another name for a
surge), and interested parties might go to considerable trouble to
substantiate the story, but almost invariably by the time outsiders
arrived the glacier was innocently composed with only its disheveled
condition showing it had been acting up.

What glaciologists long wanted to see was a surging glacier in action, but this wish was not easily fulfilled. Surging glaciers seem to
behave as children do in the game where players may move toward
the goal only when the person who is "it" is not looking at them.
They sneak ahead when his back is turned and stop abruptly when
he does an about-face.

Stories of this erratic, surreptitious characteristic of glaciers came
generally from isolated mountainous places. People there had a sim-

ple economy based on subsistence farming or grazing of animals, which could be sorely disrupted by unruly glaciers. When glaciers take to stretching out they may bury pastures and animal paths and roads, and upset in short order a region's established routine. Uncorroborated reports have told of glaciers surging so fast they overtook fleeing villagers.

The Karakoram Range of Kashmir and northern Pakistan, which is one of the most heavily glaciered regions outside the polar realms, holds some of the world's longest valley glaciers and some of the most notorious surgers. In the Karakoram and nearby ranges glaciers have a habit of breaking loose with spectacularity.

In the *Journal of Glaciology* in 1954, Ardito Desio reported on the activity of one, the Kutiàh Glacier, which occupies a valley in the upper Indus River basin. Glaciers in the area had been retreating. In the spring of 1953 inhabitants of this valley grew apprehensive as the Kutiàh, then about 12 kilometers (7.4 miles) long, began moving down upon them. Desio says: "At first, they were disturbed at the appearance of enormous masses of ice in their valley, and then terror-stricken by the continual advance of a glacier already covering a three-kilometre stretch in the Stak Valley, where there are many flourishing villages and cultivated fields. . . . From 21 March the glacier continued its downward movement and development until it occupied the whole of the Kutiàh Valley burying beneath it the luxuriant woods in the lower valley." By June 11 it had pushed forward nearly eight miles. It blocked tributary streams and dammed them up. Its average speed was 113 meters (123 yards) a day, or 10 kilometers (over six miles) in three months. The writer could cite no reason for the iceflood other than a "disturbance in the equilibrium of suspended ice masses."

Another surger, the Hasanabad Glacier in the Karakoram Range, had had its position recorded a number of times before its outburst. In 1889 Ahmad Ali Khan told of its snout being six miles back from a certain road crossing. In 1892, when Lord Conway surveyed the area, he showed the glacier eight miles back from the crossing, indicating it had been retreating those three years. But then Abdul Gaffar's field mapping the next year showed the glacier had advanced to a point only two miles from the crossing. Various accounts of its expansion indicated it advanced those six miles in one winter and spring; that is, it spread about a mile a month. Irrigation channels were rendered useless and fields could not be planted. And in 1903 the Hasanabad Glacier again advanced something like six miles to "a day's march" in just two and a half months. Then the glacier stopped and wasted, but in about forty years it retreated only one third of a mile. How much faster was its advance than its retreat!

When stories of the Hasanabad Glacier reached the outside few believed its stride could have been so rapid, but more recent glacier advances have shown that it could well have behaved as reported.

The Yengutz Glacier in the Hunza Valley is another speedster of that area. At its peak speed it must have moved forward about three miles in eight days! Its appearance and position had been described numerous times over the years by surveying teams and interested persons. A villager who had seen it surge around 1903 told his recollections in 1930 to a Captain Berkeley, who was stationed in the Gilgit garrison, and this was repeated in 1934 in an address by Professor Kenneth Mason to the Royal Geographical Society, as follows:

One day when the crops were about a hand's breadth high [i.e., May] we noticed that the water in the irrigation channels was very muddy and was coming down in greater quantity than usual. We went up the nullah to see what had happened and saw the glacier advancing. It came down, like a snake, quite steadily: we could see it moving. There was no noise. At the same time water and mud gushed out from the ice while it was still advancing and flooded our polo ground and some fields. When an obstruction got in the way the ice went round it at first and then overwhelmed it. The ice was not clear, but contained earth and stones. All our mills and water-channels were destroyed. The ice continued to move for eight days and eight nights and came to a stop about forty yards from the Hispar river. As soon as the ice stopped, the mud and water, which had been coming out higher up, stopped too. The ice remained down for fifteen years, during which time one man in each house remained in the village. All our cultivation was spoilt and we could not bring another water-channel to our fields while the glacier was below them. The Mir fed us. Twelve years ago [1918] the ice began to go back.

The Garumbar Glacier, another Karakoram glacier, also had its field day. It is said to have moved so fast that it overwhelmed two elderly women running before it. That story has been called an exaggeration, and yet one can visualize a glacier on a sloping mountainside, its front shattering as it is shoved ahead, and blocks of ice tumbling downslope like an avalanche of ice boulders. It may have been such ice blocks that struck down the women.

In the Pamirs of the USSR, not far from the Karakoram Range, there are more racing glaciers. The Medvezky breaks loose about every fifteen years. But when the USSR's largest glacier, the Fedchenko, started to surge it really caused a stir. It is about 77 kilometers (about 48 miles) long, and averages about 2.5 kilometers (about 1.5 miles) wide. It originates on the 22,000-foot-high Peak of Revolution in the northwest Pamirs. Normally it moves about 50 yards a year. A few years ago it picked up speed and began stepping along at a rate of about 50 yards a day! Villagers had to flee from its path.

Other reports of rapidly advancing glaciers have come from the Tien Shan of Inner Asia, in which range most glaciers have been retreating. The Mushketov Glacier is one that has drawn attention. Over a number of years it had lost 3 of its 18 kilometers by melting, and in 1943 the lower 6.4 kilometers of its shrunken remainder was dead ice which was expected to just melt away. This deteriorating glacier was hardly one to be on guard against. But then it happened. Back in the source regions the glacier had a rebirth. It began pouring outward, crushing and breaking everything ahead of it. One investigator, R. D. Zabirov, records: "All the 'dead' section of the glacier has turned into a heap of debris, girdling the end part of the glacier in the form of a swell. Blocks of ice were piled up one upon the other; some of them resembled sharp peaks, blades, wedges and steeples; there were gaps, cracks, grottoes and passages among them. There were black and dirty blocks of ice as well as clean ones. These blocks were constantly breaking into small fragments and falling down, disappearing in gaps." For sixteen years the Mushketov Glacier advanced at an average speed of 154 meters (about 500 feet) a year and in that time increased its area more than 4 square kilometers, or over 1.5 square miles.

Overenergetic glaciers have been observed in glacier areas throughout the world. It appears that wherever there are clusters of glaciers one or more occasionally starts racing ahead of the others.

The southern Andes have had their share of surgers. And the Alps, Iceland, and Scandinavia have provided abundant reports over the last few centuries of disastrous intrusions into settled communities in or near mountains by "racing glaciers" which slithered down their valleys, overspread fields and pastures, knocked down orchards, and crunched over roads and buildings.

A disquieting fact is that sudden surging is not limited to valley glaciers. Larger ice caps can flood too. The ice cap on North East Land, one of the islands of the Spitsbergen group north of Norway, is one that experienced a massive surge. As usual, the event became known only after it happened. A recheck of a map revealed that one section of the ice-cap front had advanced 13 miles between 1935 and 1938, and aerial photos showed chaotic crevassing through much of the ice. The cause: unknown. It was guessed to have been a tectonic disturbance or "some internal glacial cataclysm."

Historical chronicles of Norway cite many glacier advances that harassed and destroyed settlements. Clear accounts go back a few hundred years. One tells of the Åbrekkebreen, an outlet glacier of another ice cap, the Jostedalsbreen. Around 1700, before its surge, the Olden Valley into which it drained was clothed in forest and pasture and was peacefully inhabited. Then the glacier began its abnormal thrust. It clogged its own valley to the rim and drooped over the ledge of a high hanging valley overlooking the lower main valley. Below it were the Åbrekke and the Tungøyane farms. Raging torrents gushed down the ledge from the accumulating ice. They gullied the grain fields and covered cultivated farmland with ruinous outpourings of boulders and gravel, destroying the farms. Tax records show that before the glacier's rampage the Tungøyane farm had three families with three houses, 38 head of cattle, and a grain crop of 28 tønner (110 bushels). In 1728 these farmers moved their houses to what they believed was a safe location, but in 1743 the glacier "suddenly swept the new buildings away and only two persons were saved." The Åbrekke farm escaped that total devastation but it was damaged seriously enough to have its taxes reduced.

There were many such handed-down stories, but skeptics and believers alike desired to see for themselves. It is not easy, however, to find a glacier in the act of surging.

With the advent of aerial photography and the increasing number of airplane flights, surveillance of glaciers greatly improved. Surges that would have gone entirely unrecorded before were spied, and it became known that there was more surging going on than had been suspected. Surges are not rare occurrences—merely infrequent—but of such scattered distribution and short duration that it is difficult to find one taking place.

Not all glacierized areas can be watched closely but an eye is kept on some. Occasionally some glaciers that have surged come to light in aerial photographs, either by their unnatural, turgescent growth between the times when photos were taken or by their disturbed appearance—the curving surge lines and the crinkled look of deeply crevassed ice in torrent.

If a pilot or air passenger is familiar with the terrain he flies over he might spot a glacier's dramatic change right away. Then if time and funds permit, and if the region is penetrable, a ground party may be sent to investigate, but too often by that time the surge has halted and only the aftereffects can be seen.

In recent years, due to better means of transportation in rough country and the fact that scientists are working in formerly inaccessible areas in considerable numbers, some surges have been observed by workers in the field.

Two of the big outlet glaciers of Iceland's largest ice body, the Vatnajökull ice cap, which is over 3,000 square miles in area and over 3,000 feet thick, made catastrophic advances in 1964 and they fortunately were observed by the Iceland Glaciological Society and studied from the air and on the ground. One, the Bruarjökull, pushed its entire front over the relatively flat inland plateau. It was reached in January 1964, and even then in winter, when ice is stiff and slow, it was advancing at a rate of one meter, or about a yard, per hour. Iceland is a complex island—about 11 per cent ice-covered, with mountains, hot springs and geysers, and over 100 volcanoes—and so who can say what subglacial force gave those racing glaciers their impetus? Live volcanoes burn even under the Vatnajökull ice cap.

Another catastrophic advance from an ice cap was found to have taken place in the empty, remote northwestern reaches of Canada's Ellesmere Island at over 80° N on the edge of the Arctic Ocean. It was seen only from the air. The Otto Fiord Glacier rushed out from one of the most extensive, unbroken areas of land ice in the Arctic. There were no features of civilization in its way so no damage was done.

Along Greenland's fiords outlet glaciers can be seen tumbling into the ocean sometimes at a rate of 100 feet a day.

Alaska and western Canada have become one exciting focal region of interest because of the number of surging glaciers that have lately been found there. Glacier specialists of the Northwest, including Arthur Harrison of the University of Washington, and Austin Post of the U. S. Geological Survey in Tacoma, Washington, have been photographing and keeping track of glaciers in that area.

Certain surgers have attracted special attention. The Bering Glacier for one. It is the largest single glacier in North America, and when it surged it did so along a 26-mile-wide front! This sprawling mass of ice lies 50 miles east of Cordova on Alaska's south-central coast. It is about 127 miles long and 2,250 square miles in area, larger than Delaware. Between 1963 and 1966 it advanced 1,000 feet along most of its width, but about 4,000 feet in one location. For 60 miles above its terminus it was severely crevassed as a result of abnormally rapid movement. The volume of ice involved in its surge was at least 20 times greater than in any known valley-glacier surge.

The Walsh Glacier, which crosses the Yukon-Alaska border at latitude 61° N, had been stagnant and heavily covered with ablation moraine (rocky material left overlying the glacier as top ice melted off) for forty years or more. Then in 1960 or 1961 it began to quicken its pace and covered over six miles in a sustained surge that lasted remarkably long—about four years.

On North America's tallest mountain, Mount McKinley, glaciers surge. The mighty Muldrow—39 miles long—is known to have surged at least four times. One surge went four miles in a few months. Billions of cubic feet of ice came down from its upper section in a wave that moved 48 feet an hour. Normally it would be gliding gracefully down the mountain's north slope, but mountain climbers who came upon it while it was surging say it was grinding and booming loudly and sending up sharp peaks that showed the bright blue of its inner ice, which otherwise lies hidden under its dark, weather-beaten exterior.

The Black Rapids Glacier in central Alaska could not help having an audience. In 1936 it made its famous strike toward the Richardson Highway, the only all-weather, north-south road connecting Fairbanks and the Alaska Highway in the interior with the ports of Anchorage and Valdez in the milder south.

Near that highway in the Alaska Range stood the Rapids Roadhouse, a hunting and fishing lodge. A caretaker family was living there during the quiet, desolate winter. Throughout the fall they had been hearing rumbling sounds from the mountains, but they took them to be earthquakes, for the earth would shake and the building would jiggle. As the months darkened into winter the rumbling sounds grew louder and the quaking more continuous and intense. Then one morning in December the caretaker's wife again heard and felt the rumbling. It was especially close and coming from a definite direction, a valley opening in the mountains that stood along the highway. She took binoculars, focused them on that valley—and screamed!

There a gigantic wall of ice leaned forward, jumbled and cracking and casting blocks of fresh blue ice into the valley ahead of it. It was over a mile wide and up to 300 feet high. Through the binoculars the family watched sharp house-size chunks of glacier crash from the ice front and smash to the ground with a terrible roar. They knew this was the Black Rapids Glacier which should have been two miles farther back in the mountains. The glacier was bearing down rapidly on the highway and lodge and its steady approach was watched day by day with suspense. Its speed averaged 115 feet a day, and at one time it was clocked at 220 feet a day. In one three-month period it traveled three miles. It ground to a halt in mid-February just a half mile from the lodge. It continues to bear watching as it has surged several times before.

Steele Glacier, west of Whitehorse in Canada's Yukon Territory, also made headlines. It was roughly the size of Manhattan Island and about 1,000 feet thick, and presented a normal, placid appearance as it lay on the north slope of 16,644-foot Mount Steele. Then one July day in 1966, Philip Upton, chief pilot of the Icefield Ranges Research Project, was flying over the mountains as he often did from the project's base at Kluane Lake on the Alaska Highway. He looked down, and he looked again. The Steele appeared startlingly abnormal. It was cracked and shattered into sharp pinnacles the size of cathedrals.

Three days later members of the research expedition, which was sponsored jointly by the American Geographical Society and the Arctic Institute of North America, reached the glacier by helicopter. The Icefield Rangers dubbed this "the galloping glacier" and the name caught on as a new term for a surging glacier. This formerly sluggish stream of ice, which was over a mile wide and weighed in all an estimated one and a half billion tons, flooded at a rate of two feet an hour. The Steele overrode its lateral moraines, sheared sides of mountains and cut off tributary glaciers, dammed rivers and created new lakes, and, according to reports of the American Geographical Society, generated ice caverns large enough to dock the largest ocean liner in. Its surge was pulsating rather than steady. It poured ahead about 15 miles in less than a year.

The Steele was the first American glacier that scientists could watch close up for any length of time while it was in full surge. Some thirty top Canadian and United States glaciologists hurried hopefully to see what was perhaps for some a once-in-a-lifetime show. Tourists in the area, catching the contagious excitement, chartered planes to see it too. Mount Steele, one of the high peaks of the St. Elias Range, is visible from the Alaska Highway, and bus tours stop to point it out along with adjacent Mount Logan, Canada's highest peak. No towns were in the Steele's path fortunately. The nearest one, little Burwash Landing with only 200 inhabitants, was 50 miles away.

Sometimes the racing Steele sounded as loud as an express train. When ice cliffs split and the crumpling front was thrown forward the noise was deafening. But, on the other hand, L. A. Bayrock, reporting for the Boreal Institute of the University of Alberta, said his group found the glacier flowed quietly during the days they were there, and that the only sounds heard were the occasional dropping of pebbles or pieces of ice. He said the ice was twisted and looked as though chunks were going to crash any moment, but they did not. He was amazed that the glacier flowed so quietly and that the towering pinnacles did not collapse though moving along a couple feet an hour.

Obviously even in chaos the ice remains cohesively strong. And secretive. When in a state of surging, when action is speeded up and glacial processes should be more easily analyzed, this glacier—like all glaciers, surging or not—kept its secrets in deep freeze.

Walter Wood, leader of the Icefield Ranges Research Project, said, after a long on-the-spot study of this galloping glacier, that the observers still could not be sure whether the entire glacier was sliding along on its base or whether just the upper layer of ice was sliding along over the bottom one. And when asked about the causes and effects of glacier surging he frankly admitted, "Observation is all we have to go on."

Professor Samuel Collins, who had come from the Rochester Institute of Technology, honestly confessed, as quoted in *Time*, "We just don't know anything about the action and reactions of glaciers. That's why we're here."

Glacier watchers eagerly wait for other glaciers to surge. They have their eyes on some they think might, and on others whose surges still have not run their course. Motion adds interest to anything, and so a glacier too is more interesting when it shows movement than when it stands still. But the scientist's main reason for watching a surge in progress is the hope of learning how surges take place and why they happen.

Some glaciers that lie at the foot of precipitous slopes, as in the Himalayas, receive heavy avalanches of snow at times and may surge as a result. But that explanation could account for only a few known surges. Exceptionally heavy snowfall on a glacier for a period of time might also be the cause, but in no case has a definite relationship been demonstrated between heavy snows and surging. Still, surging has something to do with the carrying away of accumulated ice. Steepness of the glacier's bed has been mentioned as a factor in allowing ice to flood, but it cannot be the main cause or iceflooding of this sort would be more commonplace than it is.

No causative link has been found between surging glaciers and the atmospheric-climatic forces above them or the tectonic forces operating below them inside the earth, although it has been theorized that there might be a relationship between surging and either of those forces.

Climate is the main regulator of a normal glacier's changing rate of "speed" through its effect on the budget of snow accumulation and on the ice's temperature and rigidity, but the climatic factor cannot be proven to be the cause of any specific surge. Configuration of the glacier's bed may also be a contributing factor, for it helps determine how deep the ice can accumulate before moving.

It is thought by some glaciologists that surges might be started by earthquakes, and attempts have been made to show that surges resulted from quakes of considerable force in their vicinity, but this hypothesis has not held up. Sometimes a glacier may surge ahead after an earthquake, but then again quakes occur and affected glaciers just sit tight. And most surges occur without any big quake to induce them, as far as we know. There are so many glaciers in earthquake regions that if quakes were the main cause many more glaciers should be surging. Perhaps some quakes that have been felt in connection with a surge were caused by the ice itself moving rather than the bedrock.

When a sudden massive transfer of ice to the front of a glacier cannot be accounted for directly on the basis of climatic variation (even allowing for the time lag between snowfalls and ice formation and movement) and when no crustal rock movement that might have quickened ice flow has been observed or recorded seismographically, then we are left with the impression that some clandestine, internal change in the glaciers themselves makes them surge, and we are once again up against that sphinx-like blank wall.

Mark Meier of the U. S. Geological Survey says this about surges: "They do not seem to be caused by earthquakes as had been believed earlier, nor by climatic changes; rather, they appear to be related to some occurrence within the glacier. Knowledge of galloping glaciers is important to an understanding of the nature and extent of future movement." Meier says that further investigations of fast-moving valley glaciers may lend support to the theory that the triggering of a worldwide glacial stage begins with surge-like pulses of activity in the Antarctic ice cap. He suggests:

A surge of the Antarctic ice cap might spread "white ice" over a large part of the Antarctic Ocean; this would cause cooling all over the world, including the Northern Hemisphere, according to this ice age theory. While there is not enough ice cover involved in the northwestern glaciers [of North America] to be an ice age triggering agent, nonetheless, studies of their characteristics could shed light on the possibilities of Antarctic surges.*

The Glaciology Panel of the Committee on Polar Research of the National Academy of Sciences has written the following in one of its reports:

Perhaps the greatest problem remaining to be solved in the field of glacier dynamics is the mechanics of the sliding of the glacier in its bed and, in particular, the cause of glacier surges. . . . It has been postulated that our great ice sheets also can surge, and, when they do, an ice age starts.

We have seen that valley glaciers can surge in all parts of the world, not just in frigid, polar wastelands. They overrun buildings, farms,

* This theory is discussed in Chapter 7.

THOSE GALLOPING GLACIERS 177

towns, and roads, destroying man-made works, and even cause deaths. Not only the small ones break loose in unpredictable rampages. The largest valley glaciers in the world do, including the Fedchenko, largest in Asia, and the still-larger Bering in North America. And we know that even larger ice caps surge too. They can suddenly expand or burst out through outlet glaciers which escape through confining mountain rims. Among the ice caps that are known to have surged are continental Europe's largest one, the Jostedalsbreen in Norway; Iceland's largest, the Vatnajökull; the great Juneau Icefield on the mountains above Alaska's panhandle; and ice caps of North East Land and Ellesmere Island.

We think in still larger dimensions. Even Greenland, a continental ice sheet and the world's second-largest ice mass, sends surging distributaries outward to the sea cramming it with icebergs. Surely much surging must occur in the still-unexplored complex ice sheet of Antarctica.

Although most surges are short-lived we have seen that some sustain themselves for years and travel far. After a surge has played itself out the glacier returns to normal ground speed. Usually the added appendage simply comes to rest, gradually deteriorates, and melts away. But the time needed for the appendage to disappear is many times longer than the surge's duration. In other words, ice spreads in these cases much faster than it retreats. A glacier may surge time and time again, and one surge may override old, rotten ice left from the last surge.

All this makes one think how it might be possible for the ice to reassert its commanding influence of the past. It retreats slowly under normal conditions. And it can expand rapidly for no observable reason. Might it not have been possible for the great ice sheets of the past to surge and surge again as our smaller glaciers do now? Perhaps ice caps and ice sheets at the start of the Ice Age and at the start of each glacial stage were not as slow in spreading as established opinion says they were.

If an ice cap like the one on North East Land can advance 13 miles in three years, then in 99 years one could conceivably have spread 429 miles, the distance from the southern tip of Hudson Bay to Ottawa, or from the city of Quebec to New York City. And if a glacier surged 200 feet a day, a rate at which some modern surgers have been clocked, and if it could keep up that pace, it could advance nearly 14 miles in only *one* year—over three times as fast! But glaciers probably cannot keep galloping continuously any more than a horse can and there would have been rest periods. Knowing the capabilities of surgers, however, makes braver those who argue that the ice did not take hundreds of thousands of years, or even tens of thousands, to overspread the continent. One cannot help speculating about what could happen if many glaciers, if the whole army of glaciers, speeded up its cadence. If many ice caps were surging and the ice-free lands between them were being iced over from many centers, the surges might be encouraged by local climatic conditions to continue, even to accelerate. Even when surging stopped, the newly spread ice would not disappear as fast as it grew, judging from modern surging glaciers, but would hold its positions long after the momentum ceased.

Surging now goes on around the world, and where glaciologists look for it they are finding it more and more. The U. S. Geological Survey states, "Aerial reconnaissance studies of glaciers in northwestern North America have revealed a spectacular number of surges." And we feel an icy chill around the backs of our necks.

Sometimes the Ice Age seems not very far away.

CHAPTER 15

WE AREN'T OUT OF THIS THING YET

Tippehatchee, Vermont

Slide representing a glacier.

The unprecedented cold weather of this summer has produced a condition that has not yet been satisfactorily explained. There is a report that a wall of ice is moving southward across these counties. The disruption of communications by the cold wave now crossing the country has rendered exact information difficult, but little credence is given to the rumor that the ice had pushed the Cathedral of Montreal as far as St. Albans, Vermont. . . .

TELEGRAPH BOY:

. . . They say there's a wall of ice moving down from the North, that's what they say. We can't get Boston by telegraph, and they're burning pianos in Hartford. . . . It moves everything in front of it, churches and post offices and city halls. I live in Brooklyn myself.

THORNTON WILDER, *The Skin of Our Teeth*

ANY REALIST knows the Ice Age will not return in our lifetime. We feel quite safe from it. Yet, whenever a cold wave pierces our winter defenses, or voluminous snows close in on us, or a summer vacation is spoiled by too-cool weather, a buried thought rising more from insecurity and emotion than from reason may seep up into our consciousness: "Is this a sign the Ice Age is coming back?" Later when the weather ameliorates the silly thought is dismissed. We have been assured it would take thousands of years (surging glaciers notwithstanding) for ice sheets to re-form and become large and mobile enough to threaten areas where there are now large population centers. And it is mainly ice sheets—not just high snowdrifts or isolated glaciers or the North Wind—that symbolize the Ice Age threat. We picture a weather-beaten, invincible wall of solid ice glaring down at us from somewhere in the North. On it comes, like a champion football team on the march—grinding out yardage like a machine, lunging forward with increasing momentum despite an occasional loss, its firm offensive line powering ahead with the practiced drive of a many-time winner.

That ice wall you and I shall never see. Nevertheless, if someone wanted to give us ice-age heebie-jeebies he could at least stir up some apprehension. He could say, "Look. We wouldn't have to wait all those thousands of years to feel the effect of a new ice age. Long before there are ice sheets up north, people down here are going to know there has been a change in climate." And he would be right. He could say, "Everything has to have its beginning, and for all we know maybe the new ice age began a long time ago and we just don't know it yet." And he could be right. Maybe the whistle signaling the end of the deglaciating time-out period blew long before anyone began listening for it and the ice has put its team back in action, and maybe its big drive will start at any time. Some watchful scouts may suspect this but none can say so for certain because the field is immense and the maneuvers are hard to analyze. The ice could be expected to do some feinting and have some trial runs before launching another all-out charge, but maybe it has already done that too. Although we have been through a warm interlude the clock may still be running for the Wisconsin Stage of the Ice Age. The ice team is plainly on the field. Its back is near the goal post but that never means the game is over.

The climatic balance is unsteady and climatic trends reverse every so often. The person who contends the Ice Age is on the way back can build a fairly convincing case. He can quote sketchy statistics that suggest short-term cooling occurred in some areas in recent decades (although the long-term trend since the Ice Age has been a warming one); and he can point to the fact that there have been several pronounced cold periods within historic times.

Following the arbitrarily defined close of the past Ice Age there was a long warm period called the Climatic Optimum when temperatures were significantly higher than now. It lasted from about 5000 or 6000 B.C. to about 3000 B.C. After that there were shorter-term, down-and-up fluctuations of temperature but never again has the world been as warm as it was during that post-glacial "summer" accompanying the final recession of the ice sheets. Europe gives us our earliest, most complete post-glacial climate descriptions, and among them are accounts of varying periods of warmth and cold, dryness and wetness.

One temporary backsliding to cold conditions came about 900 to 450 B.C. during the early Iron Age in Europe. Glaciers of the Alps re-expanded and closed many passes which would not reopen until about A.D. 700. Raininess and storminess became more prevalent. Lakes rose and bogs increased on the landscape of northern Europe. In America most of the glaciers now existing in the Rockies south of 50° N formed or re-formed during this cold time or just before it when temperatures began falling.

But warmth began to come back and by Roman times climate was not far different from that of today. The years between A.D. 400 and 1200 were characterized by higher temperatures and drier weather, and the Atlantic and the North Sea were less stormy. The peak of this genial period was probably about 800 to 1000.

During this warm period leading into the Early Middle Ages forests in Europe grew higher in the mountains and farms spread farther up the hillsides. Vineyards on the continent could be cultivated farther north than now; even in England high-quality wines were produced. Arctic pack ice had melted far back, so drift ice in the seas near Iceland and Greenland south of 70° N was infrequent in the 800's and 900's and absent between 1020 and 1200. Shortly before 1000 the Vikings were exploring westward from Norway across friendlier seas than today's seafarers encounter. Iceland was settled in the ninth century, and in the 980's Norsemen colonized southwestern Greenland.

In the summer of 985 over 300 people with their livestock sailed up the fiords of southwestern Greenland to the grassy shores at their protected heads, spots Eric the Red had previously selected for them. The settlers, their descendants, and new immigrants did well on this land which we now consider forbidding. They were not just hanging on. While the environment was favorable the Greenland colonists numbered four or five thousand and maintained as many as 280 farms. They raised horses, sheep, goats, pigs, and cattle, and found products that had a ready market in Europe—walrus ivory and hides, sealskins and seal oil, white falcons prized by medieval sportsmen, polar-bear cubs, and fluffy down from eider ducks. Regular trade was carried on with the homeland. At one time the colonists had over a dozen Christian churches and their own bishop. No trees grew in Greenland even then, so driftwood was used and occasionally logging expeditions were sent to the North American continent.

The Norse colonies on Greenland had a sort of permanence in spite of their fringe locations. They lasted about 500 years. (Consider that Europeans settling the United States started their first colony at Jamestown in 1607, not even 400 years ago.) Norsemen were still in Greenland when Columbus landed in America but by then only a few remaining families were struggling to save their farms. Around the year 1500 for some unknown reason the last of the inhabitants died or disappeared. Clothing in the graves, preserved in frozen ground, tells that there was contact with the Old World even toward the end of the last colony's existence, for style details in the dresses, cloaks, and hoods followed fashion trends in Europe. But ships came less and less frequently until finally one arrived and found no more Norsemen, just deserted farms, crumbling buildings, and graves.

There is endless speculation about why the colonies declined after a long period of success. The cause might have been epidemics, battles with hostile Eskimos or assimilation by them, increasing malnutrition, or inability to build or buy ships. But deteriorating climate undoubtedly figured in the colonies' collapse, for toward the end of their existence the warm episode had passed and the Little Ice Age was about to begin. By 1200 the southward advance of Arctic sea ice was noticeable. Shipping lanes from Greenland to the European continent were stormier and more ice-clogged again, so fewer ships traveled the old lanes and isolation increased. Crops and livestock did not fare as well. The ice cap, hunched just over the mountain rim, may have exerted stronger influence on the weather. As time went on, graves were dug shallower because of increasing frost in the ground and perhaps because survivors became weaker. No one knows what happened to the last colonists.

Europe's climate had been deteriorating also. The Black Death which first appeared in the fourteenth century is probably blamed too much for the abandonment of farms and whole villages that took place then and later. Falling temperatures and resulting harvest failures were part of the cause. Cooler weather introduced the Little Ice Age, which lasted from about 1550 to 1850 and brought the greatest extensions of ice on land and sea since the Ice Age proper. Then valley glaciers grew appreciably and in many mountains of the world, particularly in middle latitudes, innumerable glaciers re-formed where none had been since the close of the Big Ice Age. People studying old maps during that cold period and afterward were amazed to see that in places where a glacier had lain for as long as anyone remembered the mapmaker of a much earlier day had drawn farms, orchards, buildings, and roads. Obviously the glacier had grown, driven away the inhabitants, ingested the cultural features, and taken over the land. Glacier renascence was a common occurrence throughout Europe, North America, and mountainous areas elsewhere. Glaciers grew in the mountains and came down threateningly and destructively into lower elevations, obstructing regular traffic through passes, some of which remain closed today. It is thought that permanent year-round snows may have covered northern Canada during part of the Little Ice Age.

Alarm about the climate's change led to the establishment of Europe's first weather stations, which in the mid-1600's began using instruments and keeping records.

American historical climatic notes and recollections do not go back as far as those of Europe, but even since the Colonial Period began some abnormally cold spells stand out. We shall look at the more memorable ones to point up the fact that the boreal whip has been coming down upon us every now and then, and also to show how the scourge of intensifying snow and cold might be felt should the Ice Age start to return. The Ice Age spirit keeps haunting his former realm, letting us feel the ghost of his old presence. Sniping warfare between Warmth and Cold continues. We are not out of this thing yet.

Temperature readings unfortunately are absent from early Colonial weather accounts because thermometers were new instruments and in short supply. In America one was used for meteorological purposes for the first time in the winter of 1717–18. It was owned by a Philadelphia doctor whose interest in weather may have been whetted by an event of the previous winter—the Great Snow of 1717, when snow fell from February 27 to March 7 in four storms paralyzing eastern New England. The first half of the eighteenth century gave New England several severe winters with temperatures low enough to freeze over New York Harbor in 1704–05, 1719–20, and 1732–33. But the winter of 1740–41 hit the colonies harder than any of those. Drifting snow closed roads in New England and the Middle Colonies; and rivers, inland waterways, and even salt-water channels froze over. On fifteen consecutive Sundays people came in sleighs from Boston's harbor islands to the mainland for church services.

The storm of January 1772 is known as the Washington and Jefferson Snowstorm. George Washington watched it from his Mount Vernon home and in his diary wrote of snow being up to the breast of a tall horse everywhere, and of people being shut up for ten or twelve days "by the deepest snow which I suppose the oldest living ever remembers to have seen in this country." And Thomas Jefferson and his bride, returning in this storm from their honeymoon, had to abandon their carriage and ride horseback over a mountain trail to reach his home at Monticello.

There was the Hard Winter of 1780 when General Washington's troops were in the New Jersey hills watching the British stationed in New York, and then the cold came and froze the water between them. Sleighs and pedestrians went from Staten Island to Manhattan and crossed the Narrows to Brooklyn, and even large cannon were pulled across the harbor ice. Deserting Hessian soldiers and others crossed from Long Island over the Sound to Connecticut. In January every port along the North Atlantic coast was shut. Most roads were closed too. Wood supplies were critically low, water was scarce because there were no thaws, and mills stood still. All colonies, even Georgia, were affected by that winter's cold, and from interior posts came reports of deep snow, and of severe cold and much suffering from Detroit to New Orleans.

There were cool summers as well as cold winters in early American history, the most famous of all being the summer of 1816, the year mentioned earlier as Eighteen-Hundred-and-Froze-to-Death. A warm spring in New England had been misleading. Salem already had a temperature of 101° on May 23. But one early June night the temperature dropped from 92° to 42° and it snowed. Fields were stiff with frost and tree leaves turned black. Cold weather followed, so cold that outdoor workers had to wear overcoats and mittens, and corn was killed. Vegetables had to be replanted. Through the summer there came a series of severe cold waves, some reaching into the

South and each destroying a good part of New England's crops. Frosts occurred in July and August even in Pennsylvania and New Jersey. Through the whole summer central and northern New England had just two stretches of over three weeks without frost, and crops suffered badly. Fear of famine gripped the area.

As America became settled farther west the weather story becomes too encyclopedic to relate here. The West had its hard winters too. After the Little Ice Age ended, cold loosened its grip somewhat but occasionally threw an exceptionally strong punch as a reminder that it still had the capability.

Like the Blizzard of 1888. It hit eastern United States on March 11 after baseball spring training had started and paralyzed a ten-state area in the Northeast. There was frost to the Gulf Coast and heavy snow even in Kentucky. The Hudson River valley had drifts sixty feet high. Houses and even three-story buildings in some towns were completely covered. In Baltimore the storm blew water out of the harbor and steamers sat in the mud.

We can recall climatic incidents in more recent years that keep Ice Age thoughts alive. Moderns are too unromantic to commemorate outstanding winters with picturesque names as old-timers did, but the winter of 1962–63 should have been given a fitting appellation. It broke all modern records for persistent cold in Europe and for snowiness in Britain. During the worst of it practically nothing mechanical moved over Europe's landscape, not trucks or trains or barges. And when they did move travel was difficult due to ice and drifts. An Austrian train going from Vienna to Paris got lost, and a Yugoslav train had its last five cars blown off. Almost every river froze, including the Thames, Danube, and Rhine; and Rome's Tiber froze for the first time in 500 years. Factories shut down. People were found frozen stiff not only away from buildings but some even at their very door. Half the birds in Europe died. *Life* magazine wrote: "It was as though—terrifying thought—a new Ice Age was at hand."

Britain had a bad time. Londoners plodded through a foot of snow for two months. The moors looked like the Greenland ice cap, and one farm was isolated for sixty-six days. Swans and fish froze solid in rivers. Service stations ran out of antifreeze, coal stocks were depleted, and power was cut back. In Scotland snow was so high people walked at hedge-top level and sheep walked onto roofs. Ireland's shamrocks all but froze out, and artificial ones had to be used on St.

Patrick's Day. Pneumatic drills reportedly were used to dig carrots in England and graves in France. In the Berlin zoo crocodiles died of pneumonia and reindeer were used to pull fodder wagons to the animal cages. Schools in East Germany were closed a month. Buses instead of ferries rode among the islands of Denmark and bicyclists crossed to Sweden over an ice pavement. The Iron Curtain had to be extended out onto the ice-covered Baltic Sea. In Paris and Naples water pipes were bursting. Starving wolves ranged out of the wilderness into populated places in France, Yugoslavia, and Italy. In central Italy they ate sheep and held up cars for hours. Near Florence they ate watchdogs and held people at bay in their houses. The south of Europe took a bombardment of blizzards and intense cold. Spain lost her citrus crop. There was skiing in Barcelona and Marseilles on the Mediterranean and snowball throwing on the isle of Capri. Subtropical Sicily off southern Italy saw twenty inches of snow among her palm trees.

The United States was not spared in that terrible winter of 1962–63. Some states saw the worst December since the National Weather Service started its records. Florida's vegetable crop was virtually wiped out and the citrus crop was 80 per cent damaged. In the Mississippi River floating ice was seen as far south as Cape Girardeau, Missouri. At the end of February the Great Lakes were nearly all frozen over; even the largest ones—Superior, Huron, and Michigan— were at least 95 per cent ice-covered.

These accounts are not just to relate chilling events that happened, but to show what would be happening over and over if we began reverting to colder climate.

The United States was hit hard again in the winter of 1965–66. Out West that would be remembered as "the winter the Ice Age returned to the Rockies." (But skiing was great.) The end of January brought a brutal cold wave. Every continental state reported below-freezing temperatures on January 29. The South got its deepest snows of the century. Lifeguards on Miami beaches wore parkas and temperatures there sank to a critical 35°. It was 24° below zero at Russellville, Alabama. That winter Maryland had its worst snowstorm in a hundred years with the Army rescuing 500 stranded motorists. In Washington, D.C., all offices were closed in one storm and snowdrifts looked in second-story windows of the White House. Snowmobiles rode at second-story levels in Oswego, New York, and helicopters brought food to stranded residents of Syracuse. One February blizzard clobbered an area 1,500 miles long and 500 miles wide. *Newsweek* said this looked like "a visitation from the geologic past." In March blizzards came to the northern plains. Four transcontinental trains were stalled by drifts but fortunately had heat and food. Many communities became ghost towns. Bismarck, North Dakota, went through eleven consecutive hours of zero visibility. Cattle suffocated when barns were sealed in snow, and those in fenced areas tramped on fallen snow until it was so deep they could walk over the fences and many wandered off and died. For the first time in history the University of Minnesota in Minneapolis called off classes.

Thoreau said it long ago: "If the race had never lived through a winter, what would they think was coming?"

If the Ice Age were really returning we would have winters that would make these look like springs.

Much could be written about weather extremes and anomalies but we shall cite just one more example of an attack by the never-say-die Ice Age troops—Chicago's Great Storm.

It came without warning. Snow began falling Thursday morning, January 26, 1967, and by Friday morning there was 23 inches of snow on the city, the greatest snowfall on record there. The big city lay still, strangled. Hotels were jammed. Twenty thousand cars and 500 city buses were abandoned on the streets. Stores closed and food was hard to get. The National Guard was summoned to clear the snow but only ten of the 350 men called could report because transportation was halted. As the city was trying to clear the streets more snow fell on Sunday—and on Tuesday. Panicky shoppers stocked up on groceries and the shelves emptied. Heating oil supplies sank. Sunday still another storm roared through the city. Clearing went on. Some snow was dumped in the rivers. Michigan and Wisconsin sent snow-removing equipment. Total accumulation now was 37 inches. Where cars were buried plows could not operate and some cars stayed snowbound through February.

In Chicago's large railroad yards switches were kept open by jets of burning gas, but the yards had to be cleared and there was no place to put the removed snow. Empty gondola cars were the only place to pile it, and when these were filled what was to be done with them? They were attached to trains going south. Surprised train watchers along the route saw loads of melting snow go by. The Rock Island

Line's snow-filled cars were sent to Fort Worth, Texas, and the Illinois Central's to Louisiana and Mississippi.

In the midst of that snow crisis the Weather Bureau stated that Chicago's storm had been so tremendous that another like it was not expected for at least a hundred years. But only the next day Chicagoans watched an ominous "low" coming east from the Rockies that had the very same pattern as the one that brought the knockout blow, and Mayor Daley asked them to "say a few prayers so we don't get it." Meteorologists then had to concede that the potential for a duplicate smothering storm was not a century away but already on the weather map.

Chicago received something like 24 million tons of snow in the series of storms that came one atop the other. And the calamity could not be foreseen.

A philosophical friend of mine has a saying: "Great storms come unannounced." Maybe ice ages do too.

CHAPTER 16

SPECULATING ABOUT THE CLIMATIC FUTURE

It must be admitted that forecasting and especially long-range forecasting still falls short of what might be regarded as an attainable goal—although without more basic knowledge of the atmosphere even the definition of an attainable goal is difficult!

WORLD METEOROLOGICAL ORGANIZATION

HUMANS migrate in response to climatic changes, even as animals and plants do, moving from unfavorable to favorable locations if a course lies open. However, since we respond more quickly than unintelligent life forms, we can move thousands of miles overnight if we wish. Or if a course does not lie open we have ways of insulating ourselves from temperature extremes and of dealing with too much or too little moisture. If climate took a definite and severe cold turn many of us would migrate to warmer regions. Or if we were "rooted" in places that were becoming inhospitably cold (or hot or dry or wet) we would artificially revamp our environment as much as possible. That is not easy. Braving the elements is sometimes foolhardy and can result in casualties, financial and physical.

Climatic changes do take place gradually, as we have said, but not always as imperceptibly slowly as was formerly believed. It used to be taught that our existing climate was a rather settled condition for as far ahead as we could see, and that if occasional sharp deviations occurred they would soon right themselves and climate would return to "normal." Archeologists and ecologists have lately been learning that certain local climatically induced environmental changes in the not too distant past happened fairly rapidly. For example: forests thinning, deserts expanding, water levels dropping or rising, rates of erosion accelerating or slowing, floods remodeling valleys, and so on.

Some of climate's "sudden" changes may indeed be only temporary extremes but while they last they are quite real, and may even be fatal to certain plants and animals. Sometimes they alter the complexion and life of an area permanently. In limited regions at least, abrupt climatic changes can noticeably transform the geographical environment and human occupation in a relatively short time.

A change of climate is the suspected cause of the decline of a number of early civilizations. In the Old World climatic upsets kept tribes and armies on the move throughout history, vying for the more productive lands and for water, and in the Americas Indians moved from one territory to another in response to increasing dryness and other climatic factors. During the Little Ice Age large areas in Scandinavia seem to have been abandoned in just one generation and disasters to human settlement in the Alps were many.

Earlier we mentioned the unusually cold and snowy weather that plagued New England during the Colonial Period and the early nineteenth century. The notoriously cold year of 1816 may have been the straw that broke the farmer's back, for there was a noticeable exodus from New England and a reluctance among new settlers to go there after repeated agricultural failures and physical hardships over a number of years culminating in that "year without a summer."

Translating such a situation to the present, one could say that if Chicago, for example, were to be snarled regularly instead of rarely by snowstorms like its Great Storm of 1967 it might not remain the beckoning metropolis it is; and if its surrounding Corn Belt were cursed with cool summers besides, that prime agricultural heartland would not be so attractive. Workers, after time to make plans, would move elsewhere. Farmers, after too many crop disappointments, would have to practice a different kind of agriculture or move farther south, and the Corn Belt would relocate nearer to Cotton Belt latitudes. And if the Deep South were badly hurt too, many people might go on to even warmer latitudes—the Caribbean or Middle America or Hawaii.

If weather did cool off generally or if increased snowiness did become a serious handicap in the North, there would be a population shift somewhat like the migration from farms to cities that we have been witnessing. Older folks with strong home attachments and lifetime investments would tend to remain where they were except perhaps in their retirement years. But young people with new fields to conquer would be freer to go where living was more attractive and job opportunities more inviting. Maybe they would move just a few counties to the south, or from the highlands to the warmer lowlands, or from inland areas to a more moderate marine coast, or maybe they would take the big jump and go where it was really warm.

Migration of population and businesses would be encouraged by inconveniences and handicaps cited earlier in descriptions of extreme winters. From management's point of view, the cold and snow would cause increased absenteeism, delays of shipments to and from the factory or store, fuel and power shortages, and higher production costs. Bad weather hurts store sales too. From the workers' point of view, there would be wage losses because of days they could not work and difficulties in getting to their jobs or keeping to a travel itinerary. Family life would be adversely affected by such things as higher heating costs; greater inconvenience and risk in driving; closed schools; canceled parties, programs, and sports events; loss of garden plants; danger of fire from overheating; fear of not getting to a hospital or of being trapped in offices or elsewhere during snowstorms. Farmers would have such problems as a greater risk of frost, crops failing to mature, damage from sleet and freezing rain, livestock stranded in the snow, and so on. In other words, the cold-weather inconveniences and disruptions that now are troublesome only occasionally would become frequent, normal occurrences in the event of a climatic step backward toward the Ice Age.

Construction work would be hampered by more frost in the ground and heavy snowfalls. Plumbing and water mains would require greater protection. Things that have to be buried below frost level would have to be sunk deeper. Taxes would go up with rising costs of snow removal and road repair. In the country cold spells would freeze out fish in lakes and deplete flocks of certain game birds and wild animals. New plants, hardier ones, would appear to take the place of those that died out. Strange birds from northern regions would make surprise appearances. Wild animals from the country would stray into the suburbs.

Some of our local wild animals, small and large, might succumb as winters became more rigorous and habitats changed. Each time glacial climate has returned additional species of animal life have become extinct. Some of the magnificent large animals that managed to survive several stages of the Ice Age fell victim to the last

one. If the Ice Age should reach another grand climax, really driving
the human animal to more equatorial climes and upsetting his favorite habitats and the geographic equilibrium that sustains his superior-
ity, our species might be one of the next to suffer depletion of
numbers or outright extinction. Our ego rebels as we consider the
idea, but we recall what happened to other well-adjusted animals
during the Ice Age. For instance, members of the elephant family
before the Wisconsin Stage walked as strongly upon the earth as
Man does now. But with the last onslaught of cold all mastodons
and mammoths became extinct, and now only a few varieties of
elephants remain. In spite of their shrinking domain and increasing
subjugation by humans these living fossils still walk with the regal
bearing of once-kingly monarchs as they wander ever closer to that
mysterious, unlocatable graveyard where elephants in fable go quietly
and secretly to die.

Homo sapiens now swarms over the planet, breeds prolifically,
and is counted in the billions, but half of our species—those in underdeveloped areas—live perilously with one foot in death's door, with
inadequate food, foul water, and crowded, unhealthful homes. Others
—those in "overdeveloped" areas, who consider themselves more
fortunate—suffer from poisoned air, nerve-wracking noise, and tension-
building social pressures.

Everywhere we look our range is overstocked. All people in varying
degrees are competing for food, personal needs, and breathing space,
and the competition increases with fantastic speed. We humans, in
spite of our large numbers, or because of them, could be the heaviest
casualty of a recurring ice age. Our life chain is all of a single species
now. We have eliminated all other possible competing Men or near-
Men. It is us or nothing human.

But let us not carry pessimism to that extreme. A new ice age need
not push us to extinction nor even set civilization back. The opposite
might be true. It might be the very thing that could revitalize a human race that was growing too pampered, complacent, flabby, and
degenerate for its own good. Instead of annihilating us, the Old Cold
coming back might annihilate other anti-human creatures—like certain insects or rats or some unsuspected aspirant to world power—
and favor us humans as it did several times before. With each of the
glacial stages our species became more advanced, more secure, than
before. One could argue negatively that we made our advances dur-

ing the warm interglacials and merely endured the glacial periods. One could, except that we do know of the astonishing stride we took during the height of the last glaciation even in the very presence of the glaciers. A new glacial stage might, like those passed, carry humans on to greater-than-ever peaks of achievement that would make us twentieth-century people look as primitive in retrospect as Peking Man does to us now.

We do not mean to peer so far ahead, however. Here we are looking just at ways our modern world might be affected if the thermostat clicked and our living quarters began cooling off. We have anticipated how people would migrate to warmer regions. Yet that prediction is not completely fair to the hardy and competitive ones among us who do not look for the easy life. Many choose, even now, to live in cold regions just as a great many land and sea animals and birds do.

Deteriorating climate would cause people to migrate much faster if they were just *Homo* (Man) and not *sapiens* (wise). They are not helplessly exposed to the raw outdoor elements, but have many ways to make life comfortable even in brutal weather. Much of our population and many bases of production would be kept in their same latitudes by tradition, nearness to a resource base, emotional ties, inertia, investments in physical structures, interreliance of a number of enterprises, and other cultural and financial restraints. If a return of the cold resurrected old frontier self-reliance, industriousness, help-thy-neighbor attitudes, and stoical grappling with challenges, then such "backsliding" might be better termed a rebirth of progress.

Even if there were certainty that Ice Age conditions were approaching it might seem just as smart, for some time at least, to stay and adapt to a cooling climate as to move. People of advanced countries could do more than hole up in caves and sod huts as previous Ice Agers did. They could do better than central heating too. Domes and rambling roofs are on the way in. We are already enclosing whole athletic fields and shopping centers. You can cover your swimming pool with a clear dome and swim in cold weather, or enclose your whole yard. In parks and universities climatrons are being built in which any type of climate can be artificially duplicated and automatically controlled by computers which regulate temperature, humidity, atmospheric pressure, air composition, wind, and intensity and color of light, even reddening it at evening and sunrise. Daily and seasonal cycles can be programed on magnetic tape. Sample vegetation of any exotic region—rainforest, desert, tundra—can be given the climate it needs to grow anywhere the climatron is built.

"Indoor towns" are being built with all buildings connected or enclosed, and there are plans to erect transparent domes over small towns and over parts of larger cities. One is being built over America's South Pole base.

Long stretches of highway could be roofed over as weather protection. Already tunnels have been bored through mountains to make transportation safer than on roads and tracks going over passes. We could even put whole towns underground to escape bad weather.

Of course, artificial shields would not stop climate's natural course if it were on the skids. Temperatures would still drop, snows would still fall, blizzards would still howl, regardless of how cozy it was inside our anthill domes. During the initial stages of neo-glaciation a comfortable, unconcerned population might for a long time ignore a gradual build-up of ice occurring in out-of-the-way places just as long as life continued as usual in its temperature-controlled, oversized igloos. People in poorer or less-organized parts of the world, who could not construct havens of artificial climate, would suffer more. Those "have-nots" shivering on the outside might be a greater threat to security—reminiscent of the barbaric Huns or Mongol raiders—than the direct threat of cold and ice.

If nature really had a mind to bring glaciers back no government or league of nations could cancel that directive. Some groups undoubtedly would try. We are forever talking about doing something about the weather and someday we shall be able to modify it to some degree. But stop an ice age? It would be easier to move to Mars.

We are fantasying here, imagining our civilization lasting through all the years necessary for ice sheets to re-form. But say it did. Say at least that so much snow and ice lay in storage on the land that sea level dropped. Then if there were the desire and capability to destroy mushrooming ice caps they might be thwarted by practical considerations—a worn-out world's craving need for living room and resources. Treasure chests of both would be appearing on emerging ocean floors along the rims of continents. What politician would dare propose melting the glacier breeding grounds, if that were possible, thereby flooding again that invaluable fringe whose development was vital to the economy and well-being of his constituents? Already our cities are filling in their waterfronts and bays to make land for urban expansion; Far Eastern cities have "suburbs" of houseboats in their harbors; and small crowded nations are pushing back the ocean in desperation, following the example of the Netherlands, the world's most densely populated country, which has long been doing this.

A sea-level drop on a broad, gently sloping continental shelf or shallow ocean basin would expose a band of new, fairly level, fertile shoreline which could be the most valuable land in the world. Even a slight lowering of the ocean would put much offshore sea bottom within reach by simplifying the job of diking out the ocean because the water would be shallower. At the doorstep of jam-packed megalopolises would appear new beaches, new unspoiled land for parks and docks and industries and apartments and shoreline drives. This would not happen suddenly, of course, but over the years new land would gradually become available. Emerging plains would also provide virgin agricultural land, fresh supplies of construction materials, and untapped mineral reserves desperately needed in a resource-hungry world which by then would have exhausted accessible supplies of many materials.

Also, if increased rain associated with reglaciation made deserts bloom again and changed swamps to lovely lakes, that advantage too might make the glacial prelude easy to accept and hard to oppose.

Human history might conceivably endure long enough to see such things as we have described happen. Civilization blossomed while ice sheets still existed, and the first villages arose while they melted. If it took no more time to put ice sheets back than it did to melt them, some of our cities might still be on the map when new ice sheets began to consolidate. Many of our energetic cities are over a thousand years old and still youthful, and those that were in existence at the time of Christ are already two thousand years old. Rome and Athens are older than that by the better part of another millennium and Jerusalem is at least 3,500 years old. Memphis, Egypt, near Cairo, is nearly 5,500 years old.

If our civilization were to last into a new ice age our technological skills, provided they too last, would give us powerful advantages in coping with it. We might live near the ice with no great hardship in protected, covered cities, or underground, and travel outside in insulated vehicles without feeling the harsh weather. We could fly over the glacier fields, even into space.

It is interesting to speculate about the future, but the trouble is we do not know in which climatic direction we are moving.*

The cold seems ever near; and every year, every season, in some place a cold-weather record is broken. However, we should keep in mind that since weather data have been kept only a relatively short time and do not include the full range of natural fluctuations, weather records, like records of a new baseball team, are still easily surpassed. The United States Weather Service was not established until 1870, and many large American cities have kept records for only a matter of decades. If we want to keep alive the Ice Age scare we can continually find reasons to conclude that the climate is cooling by pointing to new snowfall records or to lower-than-ever temperatures for a given day or month in one place or another. But these are new extremes only to weather books' *recorded* figures. Weather records will continue to be broken everywhere, and in opposing climatic directions, for a long time to come even if climate remains stable, just because meteorological data are so incomplete.

If we stop to think, we hear just as many cases of heat records being broken as of cold ones. Climate is so good at balancing its deviations over the years and from place to place that the pluses and minuses appear to even out. It turns one way in one area and the other way in another. If your area suffers from drought, chances are that some place not far away is soaking in unusually heavy rains and floods. And when you hear of some part of the country establishing new records for cold or snow, another part will simultaneously or soon be experiencing abnormally high temperatures.

Examples of climate's contradictory nature are all about us. A few years ago the East Coast was parched with drought while the West Coast was washing away in floods. One recent New Year's, while New Orleans had its heaviest snow of the century, North Dakota had spring-like temperatures in the sixties. Locally I recall other instances of weather's counterbalancing effects, as you may in your area. Lack of snow in northern Wisconsin one year forced cancellation of winter-carnival events and at the very same time southern Wisconsin

* A reminder: "weather" is the day-to-day condition of the atmosphere, whereas "climate" is an average of weather conditions over a period of many years.

cities were smothered with snow, wondering how to get rid of it. And when Chicago staggered under its Big Snow of 1967 parts of southern Wisconsin just over the state line were nearly snow-free, having the least snow of any winter since the turn of the century. A few years ago Wisconsin had the coldest December in eighty-seven years, and some climate prognosticators used the occasion to warn the public that their prophesied cold trend was under way. Then the old pendulum swung, as it is wont to do, and to their embarrassment January and February were delightfully mild; in early February thin-ice and grass-fire warnings were issued in Madison, and on March 2 the first water-skier appeared in central Wisconsin—while southern California shivered with temperatures in the fifties. Throughout the world in like manner climate goes to one extreme in one place and to an opposite extreme somewhere else. What an overall "average" is at any time no one can possibly tell.

We northerners know how often a really numbing or snowy winter is followed by one that is mild or almost bootless. Winters are so different from one another through the middle latitudes—as are all seasons—that normalcy cannot be defined. We know climate is changing but we do not know *from* what or *to* what.

Plotting a trend is hard enough when you know where you are starting from, but we do not even know where we stand on the climatic graph, not in any concrete way. We need better numerical data to work from, but statistical coverage until now has been meager and spotty.

Looking back to the time before the use of weather instruments, one has to rely on sketchy descriptions from old literature and chronicles or ships' logs, and they give only fleeting, imprecise glimpses of the passing weather, and only in limited parts of the world. Weather and climate, except when unusual, were not usually written about by travelers and historians of old. History tells us little about climate directly. It focuses mainly on wars, catastrophes, and political leaders, as news reporting still does. The topics of weather and climate, which affect us all quite personally, are passed over lightly except for the freak or dramatic occurrence. For information of paleo-climates before the time of writing, one has only Earth itself to examine. The orientation of ancient sand dunes shows the prevailing wind direction when they formed. The way rocks disin-tegrated and the way the eroded particles were worked upon and moved about by wind, water, or ice in the past tell something of climatic conditions then. Tree rings record good growth years and bad. Old permafrost left soil patterns and ice-wedge casts in the ground. Fossil vegetation and pollen tell indirectly what the former climatic environment was. So do remains of animals and marine life whose climatic range is known. Old soil horizons speak of old temperatures and moisture conditions. These and other evidences of former climate show past conditions and trends only in a general way and only for certain places at approximate times. They too are an incomplete record. To be meaningful they must be properly dated, and dating is no small problem.

Even with the invention of calibrated weather-measuring instruments which supplied absolute data—that is, real numbers—the record was little improved for a long time. The thermometer was the first weather instrument devised, in the middle of the seventeenth century, and others came later. Years passed before instruments were in general use and then they were still not standardized. Figures from them must be used cautiously. For example, when we make comparison with old temperature readings we cannot assume that the readings were all made with instruments whose calibrations matched nor that they had comparable reliability and exposure. It makes a big difference whether a thermometer was on the ground or the roof or against a heated wall, or if it was rained or snowed upon, or partly in the sun, or at what hour it was read; and such facts we do not always know.

Commonly in the old days records were kept of temperature only, and, of course, infinitely more goes on in the spacious, turbulent atmosphere than is shown by the temperature at a given spot on the ground which is read only once or twice a day.

Gradually governments took more control over the compiling of weather statistics, stations became more numerous (but still were thinly and unevenly scattered), and the manufacture and use of instruments more standardized for equivalency and accuracy. But still there were complications.

Weather stations have shifted around over the years. Many lack continuity, and some that appear to have continuity actually do not. Of some 200 city weather stations in the United States whose records go back to 1900, over 90 per cent have had their statistics devaluated, so to speak, because of one or more moves during their

PLATE 21.   Looking out of this tunnel, which was cut by a subglacial stream.

PLATE 22. Uneven deposition of glacial drift and disruption of old drainage patterns left much of Finland covered with lakes. Lakes are common features in recently glaciated regions.

PLATE 23. A series of raised beaches and shorelines along Hudson Bay in northern Ontario illustrates how the land gradually rose as the weight of the ice sheet was removed. The land is still slowly rebounding, causing the bay to shrink.

PLATE 24. The Rhone Glacier in Switzerland as it was in the early 1800's. A hotel was nearby and sight-seers wandered up onto the glacier.

PLATE 25. The shriveled Rhone Gla a recent photograph.

PLATE 26. Glaciers in mountains around the world send to surrounding lowlands gushing streams of fresh water even in drought seasons when the flow of other streams lessens or stops.

PLATE 27.   Worthington Glacier in the Chugach Range near Valdez reaches towar
the Richardson Highway, a major highway in Alaska.

PLATE 28.   Tikke Glacier, northern British Columbia, after it surged. The highly crevassed chaotic surface and the misshapen, contorted medial moraine, characteristics of surging glaciers, are the result of rapid "flooding" of the ice. Lakes have formed where the ice dammed side valleys (lower right). This glacier is 16 miles long.

The dramatic retreat of Muir Glacier (named for John Muir) in Glacier Bay, southeastern Alaska, within two decades is seen in this series of pictures taken from the same point at approximately ten-year intervals.

PLATE 29. In the first picture, taken in 1941, Muir Glacier was vigorously advancing from the valley around the corner at the upper left, toward the lower right. Joined by its tributary, Riggs Glacier, from the upper right, it filled the valley.

PLATE 30. In 1950 the ice had retreated, leaving the vacated part of the valley an inlet of the ocean. A medial moraine separated Muir Glacier coming from the left of the dark mountain and Riggs Glacier coming from the right of it.

PLATE 31. In 1961 Muir Glacier had receded so far up the inlet to the left that it was no longer visible. Riggs Glacier barely reached tidewater. Icebergs from both glaciers drifted in the water of the encroaching ocean.

PLATE 32. Three U. S. Navy icebreakers shove a huge iceberg out of a shipping lane leading to McMurdo Station, Antarctica. It measures 800 by 200 feet and rises 80 feet above the ocean surface. As little as 10 to 15 per cent of an iceberg's volume is above water.

PLATE 33. Hikers enjoy an invigorating climb on the Svartisen, Norway's second-largest glacier, whose vulnerable surface here is darkly soiled and without fresh snowcover.

PLATE 34.   A class of mountain climbers in the Alps learns how to cross a crevasse.

PLATE 35. Every skier dreams of cutting his trail through limitless virgin powder snow in an unmarked setting such as this on Gilmour Glacier in the Cariboo Mountains of British Columbia. With the assist of aircraft many formerly inaccessible fantasylands in glaciered mountains have come within reach.

PLATE 36. In the Cariboos helicopters now transport skiers to the top of snowy mountain slopes, meet them later at the bottom, and fly them back to their lodge.

PLATE 37. Buildings and foot trails appear in most unlikely, seemingly uninhabitable places, among glaciers at the ceilings of the world.

PLATE 38.   Pleasure boating and picnicking are common pastimes at glacier lakes.

history of operation. Some cities, for instance, have moved their station from atop City Hall to the new airport on the outskirts of town which is in a filled-in marsh, conveniently level for planes but often subject to air drainage (cold air settling in low spots) and having quite a different "climate." Even most stations that have stayed in the same locations have new micro-climates. They may no longer be representative of the general local condition. Tall buildings may have grown up around them, affecting temperature and wind; or they may have become surrounded with asphalt instead of grass, or chimneys that blow out heat and smoke screens; or other changes could have substantially altered weather in their immediate vicinity.

It is a significant point that most continuous weather records throughout the world are from urban areas which have been rapidly changing while the open countryside which is more representative of natural weather has been largely skipped over. An even greater data gap exists over the oceans, and they comprise nearly three fourths of the globe's surface. Those vast weather-brewing regions—overwhelming in area and climatic influence—have contained few weather-recording points: some islands and buoys. And ships and planes have been making observations, but these are limited to main shipping and airline routes. Also, large expanses of the tropics and polar regions, both highly significant weather-wise, have been omitted or neglected. Better world coverage is being worked out. Most monitoring up to now has been done at the earth's surface, at the bottom of the atmosphere, and relatively little *in* the atmosphere at intermediate and upper levels, where most of the weather-making processes are going on.

Only in imprecise ways, then, can we see where we have been climatically and where we are. We have come a long way in improving methods of collecting information but the data are still woefully inadequate, as anyone trying to make forecasts or predict trends will tell you. Meteorologists around the world are cooperating now to improve the collecting and transmission of data.

A trend cannot be positively determined except over a long period of time—generations or centuries or longer. Climate, in the part of the world we know most about, has been fluctuating, so we cannot tell whether it is moving back toward the Ice Age or away from it.

A change in the trend, whatever it is, could occur any time. This fickleness and unpredictability of climate's direction reminds one of

the ups and downs of the stock market. The line on the graph
wavers continuously. Some stocks move up while others go down,
as temperatures at different weather stations do; and a temporary
overall decline or rise can occur even while the long-term movement
is in the opposite direction. But at least in the stock market there
are exact figures to look at, there are known regulators, and human
intelligence is running the show. With weather there is none of that.
"There is still no basic theory as to why the climate changes,"
commented Dr. R. C. Sutcliffe, director of research in the British
Meteorological Office. "We are in the same stage as pre-Newtonian
astronomy: we observe phenomena but we have not worked out laws
to explain them."

In the years ahead climatologists are bound to discover some of
those laws. With satellites taking motion pictures of clouds, register-
ing radiation balance, sounding the atmosphere in depth with
sophisticated instruments, and collecting data from automated ocean
buoys and land stations; with thousands of balloons measuring
temperature, pressure, humidity, and winds at various levels of the
atmosphere; with a much denser network of stations; and with super-
computers analyzing the data—with all this, great progress is bound
to be made. Some of these improvements are plans of the future.
Some are already in operation. But at this point in time we are still
in the dark. The science of predicting weather and climatic trends
is still in its infancy. Even with mathematical models of the atmosphere and with computers deciphering data, we will not know a trend
until it has taken place. Mathematics can tell the probability of something happening, but there is no assurance that such will necessarily
come to pass. The weather makes the data; the data do not make
the weather.

Since we are without evidence to indicate whether Earth is heading toward a new cold glacial period or a warm interglacial one, and
since contemporary climatic extremes and deviations are self-erasing
and contradictory, we are at a loss to say where we are drifting
climatically. Human interference with normal weather might alter
climate's course, and that is another factor to be reckoned with. The
only thing we know for sure is that climate is changing, because it
is always changing. And however it changes, glaciers will be affected. They will either grow or shrink. Those who feel concern for
the fate of glaciers listen sharply to climatologists for word of any

new finding that will indicate what that fate will be, while at the same time, interestingly, climatologists are carefully studying glaciers to learn from their growth or shrinkage what the climatic trend is. Glacier fluctuations are one of the best visible indicators of climatic change that scientists have.

You may be confused by predictions from the experts when you hear one say the climate will be warmer and another say it will be cooler. This alone should tell you the score. If you listen long enough you may even hear the same expert predicting cooling weather one year (while you are having cold spells) and a few years later, warming weather (when you are having warm or hot spells).

Predicting the climate to come is a gambling game that tempts many scientists. Some predictors speak with more prudence and restraint than others, and most admit honestly that they are making an educated guess. Everyone wants to be right, or preferably famous. And to be famous while one is still alive (although he be known as a false prophet later) is seemingly a much-desired satisfaction.

How can people be expected to take seriously predictions of worldwide climate a few decades or centuries from now when they still cannot get an accurate local forecast for tomorrow's weather?

As for a five-day forecast, it is even less reliable than the twenty-four-hour one—about as reliable as Grandpa's rheumatism; and a thirty-day forecast is no better than The Farmer's Almanac. Weather simply cannot be foreseen that far in advance. These forecasts give an idea of what to expect and little more.

Although one cannot put much faith in long-range forecasts they do command interest, and so we shall keep hearing them. Predictions of a colder climate are usually given more publicity than predictions of a warmer one, and are more readily believed. Instinctively we are always on guard against the cold.

It could be, we are so fearful with every slight draft from the North that blustery Cold might be leaking in some rear window that we have not noticed that quieter, less conspicuous Heat is wafting blithely through the front door.

CHAPTER 17

GLIMPSES OF AN INTERGLACIAL SUMMER?

We've had enough of summer. We need autumn chill and autumn rain.
EDITORIAL, *The New York Times*

WHEN people hear predictions that the climate will be cooling their response is almost always one of disappointment. They think of wearing heavy clothes more of the year and having shorter summer fun seasons. But when they hear predictions of warmer climate they envision longer, lovelier summers, more months of casual clothes, the vacation spirit for more of the year, iced drinks in a tree-shaded garden, water sports, swimming pools, picnics on the lawn, and camping in the green woods.

If we were choosing between the opposites, Cold and Heat, Cold would seem more frightening and Heat safer and easier to cope with. Heat seems friendlier, less aggressive, less to be on guard against. Cold is stern, demands exertion, stirs the blood, and charges one with energy, while Heat saps energy and makes one lethargic and complacent. Which then is the greater friend? As with our acquaintances and family members, the one who is stern and prodding, who disturbs our inertial serenity, is often more our friend than one who soothes and lulls and condones the easy way.

A couple of chapters ago we looked at some memorable cold and snowy spells of the past and the hazards and inconveniences associated with them, but before we start cheering for increased warmth to overtake cold we should take a good objective look at *its* effects too and see if that is really what we prefer. We do not have space for a comprehensive worldwide survey of warmer-than-average weather spells but we shall look at some from the last decade which touched us closely.

From the following examples alone, one might get the impression that heat and drought were continuous through the period described, at least in the United States. This is not the case. In the chapter "We Aren't Out of This Thing Yet" we saw that some exceptionally cold and snowy weather also occurred in some of these years. So keep in mind that the conflict between Hot and Cold, and Wet and Dry goes on all the time, and we have just tied together a series of weather incidents which the reader may remember to show what living under warmer conditions might be like.

Though drought was associated with many of the hot spells, heat and drought do not necessarily go together. The world has cold deserts and hot, steamy rainforests; and the high-sun season in most regions has more precipitation than the low-sun season. Also, during the Ice Age the greatest aridity did not always occur during interglacials. So we cannot say that if we entered an interglacial phase our climate would necessarily become drier, though it might. In a given situation a rise in temperature does increase evaporation. Then vegetation, animals, and people require more water. If precipitation does not keep pace with temperature rise, drought occurs.

The "drought" conditions discussed in this chapter were not wholly the result of reduced precipitation, or water shortage. They were in part due to poor distribution of water, and to our abnormally high consumption of water in warm weather because of our high standard of living—and comfort.

All factors considered—natural and cultural—if our climate should become warmer, the need and demand for water would most likely increase and we would be faced with drought-like conditions even if there were no actual decrease in precipitation.

The examples ahead are not meant to be evidence that the "warming" predictors are more right than the "cooling" ones, but they do counteract the scares sometimes voiced by the latter. They will remind us that Heat, which runs this world, remains a formidable adversary of Cold, and will illustrate what might happen if we trend further away from the Ice Age.

We pick up our narration after the Terrible Winter of 1962–63, which evoked dire warnings of the return of the Ice Age in the United States and Europe. Only the next summer the cry of drought was heard across the land.

By fall New England's water supplies shrank to the lowest levels in twenty-five years. Hundreds of forest and grass fires blazed in the East and Midwest, and woodlands were closed from Maine to Arkansas. Missouri communities were rationing water. Great Plains stockmen were hauling water to livestock. In northern states autumn was unusually warm; fall clothing was not selling in the stores, nor was antifreeze. By November New York City's reservoirs were down to only 19 per cent of capacity, and New Yorkers were asked not to let faucets run to get cool water, not to flush toilets except when necessary, not to take tub baths, only showers, and then to let the water run only to wet themselves and rinse. Over winter the Mississippi River was so low that shipping was only 10 per cent of its high-water tonnage.

It was hoped that 1964 would be a better year. It wasn't.

The drought in the United States was called the worst of the century. Crops on the Great Plains were threatened. But the West is used to droughts. The normally humid Northeast was suffering more. Wells were running dry. In New England farmers were buying water and hauling it in milk and fire trucks to water their cows. The Great Lakes level fell to an all-time low, so lake freighters had to carry lighter, less profitable loads to negotiate harbors and channels. Harbors had to be dredged and docks moved. Hydroelectric plants were short of water. Great Lakes resort operators and marinas were in a bad way, and some lake-side cottage owners had to walk a half mile to reach their boats. Mowed lawns surrounded piers where water used to be. Inland lakes and marshes were drying up too.

The drought went on but no one could say why. Meteorologists variously blamed it on sunspots, the jet stream, air pollution, a "warm pool of water in the middle of the Pacific that set up some kind of chain reaction," or cool water off the Eastern Seaboard.

The year 1965 brought fears of another Dust Bowl on the Plains. Over two million acres were damaged, and strong winds dropped dust as far east as Pennsylvania. Some dust storms blotted out daylight, blinded cattle, closed schools, disrupted power service and traffic, and filled hospitals with respiratory patients.

This drought was the most prolonged and severe on record for the Northeast. Now in its fifth year, it had spread from Maine and southeastern Canada to Virginia, and west to the Appalachians. Illinois, Indiana, and Missouri were feeling it and so were even Florida, Louisiana, and Alabama, some of the most humid states.

Rutgers University advised New Jersey people to use rinse water from the kitchen, laundry, and bathroom to water their trees and gardens. In New York City a restaurant could not serve water unless requested or it was fined fifty dollars. The city's upstate reservoirs were running dry. Another of New York's main sources of water, the Delaware River basin, was also in jeopardy, for it supplies water to four states, including the Philadelphia metropolitan area, which competes with New York for its water. The Delaware's flow was now so low and consumption so high that at the river's mouth not enough water flowed out to keep ocean salt water from pushing into the estuary. Philadelphia was told its water would soon be undrinkable, and food-processing and bottling companies there feared they would have to shut down.

Florida's rainfall is heavy, usually double the national average, but that state too was in trouble because of growing water demands of irrigation, agriculture, and real estate developments. The Everglades, largest of all United States swamps, was drying up. It comprises two thirds of Florida's fingertip where it dips just into the tropics and is the last refuge of certain rare plants, and the home and breeding place of many endangered birds and animals. It is mostly sawgrass marsh broken by higher wooded islands and owes its existence to the outflow of Lake Okeechobee, which ever since the Ice Age has poured south to the sea as a wide shallow river. A steady flow of fresh water through the glades is necessary to maintain the habitat and prevent intrusions of salt water; and if water level drops, island breeding sanctuaries are no longer protected by encircling water. Now the water level fell, the tall grass turned brown, trees died under the blazing sun. There was fire and smoke. Wildlife was dying, the beautiful birds flew away, and tourists who came to see them saw desolation instead.

Other countries in 1965 experienced drought also. England, known for its rain, had to restrict watering of gardens and car washing. Riviera tourist business along the Mediterranean was hurt by too-dry weather and threatening forest and brush fires. In Madrid that summer water was turned off every night. Portugal's hydroelectric power production was seriously reduced. In August the USSR was ordering millions of tons of wheat from Argentina and Canada to make up for crop failures. Northern China had another in a series of poor harvests, and in South Korea wells were drying up and drinking water was rationed from firefighting supplies. South Africa went through one of the worst dry spells anyone remembered.

We skip to the next summer. Relief must be near.

In New York, Mayor Lindsay lifted a ban on air conditioning in restaurants, apartments, offices, and stores whose cooling processes used water. Droughts, after all, do end. Only this one did not end.

July was horrendous. It made winters, even the Terrible one, look lovely. Week after week the weather was scorching from the Rockies to the Atlantic. Death rates increased sharply that month, as high as 41 per cent above normal in some places. In St. Louis the coroner said bodies were "coming in like a parade." As overworked air conditioners put a strain on the electrical supply that city cut off power on a rotational basis to conserve it, and families without air conditioning went to restaurants and crowded theaters, as they did in other cities, to escape the heat, but after being crowded awhile even these places were no longer cool. When the All-Star baseball game was played in St. Louis on July 13, 400 fans were stricken with heat prostration.

In the midst of the heat wave there was a massive power failure through much of Nebraska. Milwaukee was having weather typical of southern Illinois; there people were sleeping at night in the park along cool Lake Michigan. That July was the most violent month in Chicago's history, with seventy-two murders. The homicide chief blamed exceptionally hot weather. Chicago's water commissioner warned, "We're going to burn out the pumps," for it was impossible to handle the city's water demand and supply the suburbs. In some districts toilets could not be flushed for three days. A ban was put on lawn sprinkling, car washing, filling of swimming pools—just when people wanted this water most. Fines were up to $200.

Crops seared in the fields. Pavements buckled in the cities. Newly planted trees in Washington, D.C., shriveled. Drawbridges in New York did not close, as that long-suffering city experienced the most intense heat wave in its history with nine consecutive days over 90° and a peak of 106°.

The nostalgic vision of an ice age must have seemed as delicious as ice-cold watermelon.

August. No let-up. Temperatures hung near 100° from New England to Maryland. Pittsburgh endured two weeks of 90° temperatures, and Harrisburg had three days of 104°.

The drought had gone into its sixth year, and the whole nation was hurting. Corn through the Midwest was in a critical stage. Kansas cattlemen were selling their herds. The water table in twelve northeastern states sank to all-time lows, and eastern rivers were down to trickles. New England farmers were buying hay for dairy herds, and some were selling their farms. Throughout the country wells were being drilled by homeowners, even in cities; pastures, lawns, and golf courses were hard and brown; trees were dying, ponds disappearing, and crops withering. Increased irrigation only intensified the water shortage.

In 1967 the drought eased in New England but continued in other areas. By late summer forest and brush fires were headline news. Fires swept unchecked through forests of Oregon, Washington, Idaho, Montana, and British Columbia, and firefighters were exhausted. In that area 7,000 fires burned nearly 150,000 acres, including 5,000 acres in Montana's Glacier National Park. Forests in some areas were closed to loggers and sportsmen for six weeks. Lumber mills were paying seven times the usual rate for timber and some shut down.

Out-of-control flash fires raced through southern California's brushy hills and canyons. In October 5,000 persons fled their homes and left schools and businesses. Refugee centers were established; towns and oil fields were threatened; crops suffered thousands of dollars' worth of damage.

Cross-continent a stench hung over Florida's Everglades. All the water was gone from Everglades National Park except at the northern edge. Animals were too weak to breed. They struggled to waterholes, bogged down, and died. Birds had flown off to nest elsewhere, many of them the last of their species which could probably survive nowhere but in this preserve. Alligators were dying in the drying mud, and with no marine life to eat were turning cannibalistic. Conservationists captured some and took them to places where there was water.

That wish most of us had for a warmer interglacial-type climate, how does it look now? How pleasant would longer, hotter summers be with swimming pools that cannot be filled; with garden hoses that cannot be used to refresh the sunbather or water the lawn and garden; with air conditioning off because of restrictions or power failure; with dust storms, dying livestock, and ruined crops; with once-green forests and camp sites burned or closed? The tree you would sip your lemonade under is dying, the picnic lawn is brown and cracked, the flowers are stunted or dead, the car cannot be washed. The days are steamy, the nights suffocating and sweaty, and baths are restricted. Water pressure is low and at times there is no household water at all. You cannot swim or put your boat in the shrunken, warm, scummy lake. The amount of sickness and death caused by heat makes that caused by heart attacks from shoveling snow look insignificant.

We have seen that the inconveniences, health hazards, and disasters brought on by cold and snowy weather are no worse than those brought on by hot and dry weather. To many minds they are far less discomforting, do less damage, and are easier to cope with than the ravages of heat.

The pros and cons of warming and cooling climate have been examined not just for the fun of rating one against the other, but for a more important reason. We have reached a point in civilization where we are beginning to alter our climate—definitely on a local scale and possibly on a worldwide one. We may be doing it as handily and recklessly as we altered our landscape in the past. It behooves us to think ahead and begin deciding—if we shall have the chance to vote a preference—in which climatic direction we would like to go.

CHAPTER 18

CLIMATE MODIFICATION, LIKE IT OR NOT

It is becoming painfully apparent that the weather changer has forged mankind a key to the Pandora's box of weather, a box that is proving not always to be filled with pure blessings. . . . Weather changers would do well to look carefully and perhaps count to ten before leaping at clouds with silver linings.

D. S. HALACY, JR.

MANKIND seems destined to find ways to intentionally modify climate on a broad scale. The Man animal was already fighting the elements far back in the Ice Age with micro-climate techniques—a windbreak, a shade of branches, then a hammering of rock against rock to ignite a spark of warming fire that he could feed or extinguish as he wished. His ability to cope with the elements grew with time until in the last century it accelerated rapidly and he began altering the large macro-climate inadvertently even before he knew he could. Within only the past few years a slow insecurity began creeping over him as he realized that his agricultural and industrial revolutions, his widespread resurfacing of the terrain, his exhaust-expelling conveyances, have not only affected the daily weather he meets when he walks out the door but may be changing climate around the world.

Plans to revamp our geography that make past environment modification look like child's play are already being earnestly considered. Put them all together and see that astounding, far-reaching remodeling of our environment is anticipated. Some of the plans sound reasonable and practical. Some we would call fantastic—except that they are already on the drawing boards. Already engineers, regional planners, environmental scientists, and quacks in this and other countries are tampering with climatic mechanisms that no one understands, whether citizens of the world approve or not. Eager experimenters cannot wait. While some may turn out to be saviors, more are likely to be blunderers. There is hardly a building where the artificial heating-and-cooling ventilation system works satisfactorily, and yet some weather wizards want us to believe they are prepared to improve upon the circulation of God's atmosphere.

The World Weather Watch, established in 1968 to remedy the lack of meteorological data and facilities over most of the earth, hopes to become a global observation network and a data-processing and telecommunications system to pool weather information from land and sea around the world. As these data are amassed, the temptation to "improve" climate or modify weather will increase.

After one of the organization's recent international conferences the journal *Science* reported: "In America, weathermen have progressed from simple observation, to theoretical models, to efforts to modify certain types of weather. The same pattern seems to be developing in the World Weather Watch." And *Weather*, published by the Royal Meteorological Society, London, carried this prediction: "One of the most important long-term results of the World Weather Watch is the possibility that the increased understanding of the behaviour of the atmosphere which will result may make it possible to investigate large-scale weather and climate control."

Besides the many weather-modification experimenters in the United States there are over a hundred stations in other countries engaged in this research.

A look at some climate-modification schemes will show that scientists and planners are indeed disposed to modify our environment, ultimately on a grand scale. One can imagine how the world might be affected if climate changers are successful, how conditions could be manipulated for or against our lingering glaciers, and how the coming or going of an ice age could be greatly speeded up. Some schemes bear directly upon the fate of glaciers, others indirectly.

Cloud seeding is experimental, its results unappraisable, but it is already a common practice around the world. It is, in fact, big business. Active in this field are government agencies dealing with agriculture, the interior, defense, commerce, space, aviation, atomic energy, etc.; and professional cloud seeders are hired by power and water-supply companies, ranchers, farm associations, plantations, paper and pulp companies, ski resorts, and other enterprises in attempts to manipulate precipitation.

There is no way to accurately predict, evaluate, or control the results of cloud seeding. Meteorologists do not yet understand how even normal, unseeded clouds behave or exactly how precipitation forms. But more and more cloud seeding is being done. Some is regulated but much is not.

As geographer-climatologist David H. Miller expresses it, "The cloud-seeder . . . reminds one of an amateur chess player, boldly advancing his queen into a complicated terrain for an immediate profit or just to see what might happen."

One reason the effectiveness of cloud seeding cannot be measured is that the natural precipitation process has to be pretty far along before the seeder steps in. He cannot seed clear blue skies or just any old clouds. He needs clouds that are already precipitation-prone, the type that could normally precipitate but just might not. The situation is always such that precipitation could very well have occurred without any extra inducement. Analyses made of after-seeding precipitation itself to see what traces of seeding agents were in it showed no conclusive evidence as to whether the precipitation occurred normally or because of seeding.

In artificial cloud seeding the addition of innumerable microscopic particles that attract water vapor and serve as nuclei for moisture to collect on theoretically should stimulate precipitation, but there are myriads of such tiny nuclei in the air all the time, including dust, pollen, bacteria, smoke, and salt from ocean spray. A cloud seeder cannot know how much seeding has already been done by nature or man's activities, so he cannot determine the effect of his own seeding. Often when clouds are treated more precipitation comes from unseeded clouds than from seeded ones. And in tests conducted in certain areas where cloud seeding was methodically used, precipitation over a period of time was actually *less* than normal during times when seeding was done.

Producing precipitation is only one problem of cloud seeding; making it fall where wanted and controlling the amount are others. And though various cloud seeders claim they have been able to start precipitation none has any method for stopping it once it starts. That detail is left to good old Mother Nature. If done on a large scale cloud seeding might have side effects on other atmospheric conditions. But despite all the problems and uncertainties, experimenting has to go on, for the unknown calls to be explored.

Practical applications of cloud modification have begun.

Snowstorms are being manipulated around eastern Lake Erie and southern Lake Ontario, an area often paralyzed by heavy snows. Dry air masses evaporate considerable moisture while crossing the open lakes in winter and release it when lifted upon reaching shore. Cloud-modifying experiments are used to try to reduce the snowfall near the lakes and spread it farther inland.

In the Rockies tests to increase the snowpack are conducted. Techniques are so advanced they are automated: ground-based generators are placed at high and remote locations, and on radio command seed the air currents with the purpose of causing snow.

Lest this make glacier lovers say, "Hurray, more glaciers," we hasten to add that along with plans to increase snowfall there are other plans to decrease it. Snow-suppressing measures might be applied where people want access to mountain areas or use of areas subject to avalanches, or where they may want an amelioration of climate that could come from lessening snow and ice in their vicinity. There glaciers and near-glacial snowfields could be in danger of drying up.

Although the effectiveness of cloud modification is dubious there are some positive results or the practice would not be continued. Sometimes clouds do respond directly and visually, and fog at airports can often be cleared by seeding. So there is reason to believe we can someday modify cloud processes and precipitation in a limited way, and to some degree regulate cloud cover, suppress hail, redistribute precipitation, lessen dangers of lightning storms, prevent tornadoes, and tame hurricanes.

Remodelers of the climatic environment would do more than modify the atmosphere; they would also modify the oceans. Diverting of warm and cold currents seems a superhuman job but even such projects are contemplated.

A plan exists to blast a new canal across Central America with atomic explosives. Not a canal with locks at different levels like the Panama Canal, but a sea-level one through which waters of the Atlantic and Pacific could mingle. There might be a net flow of colder water from the slightly higher Pacific to the Atlantic, cooling the Caribbean area and the Gulf Stream, and so cooling and otherwise changing the climate of land now warmed by that current—eastern United States and western Europe.

Closing connections between oceans is also being considered. Some Russian planners want a fifty-mile dam built across shallow Bering Strait between Siberia and Alaska to keep cold currents from the Arctic Ocean out of the North Pacific. This would make the Pacific shore of eastern Siberia warmer and give the USSR more ice-free ports. But in environment modification as in anything else, for every action there is a reaction. The cold Arctic Ocean water has to move out somewhere, so if it could not exit through Bering Strait it would leave through the openings between North America and Europe. The already icy Labrador Current could become still more frigid, intensifying cold along eastern Canada, hurting agriculture there, closing ports, and carrying the cold even farther down to eastern United States.

The Russians have another version of the Bering Dam plan. Instead of just confining the cold Arctic waters, they would pump warm Pacific water northward into the Arctic Ocean using hundreds of giant nuclear-powered pumps along the dam. This would warm the Arctic Basin and improve Siberia's severely cold climate. It could also hasten the melting of Greenland's ice sheet.

Soviet engineers hold different opinions about how to use this dam. Some think it would be better to pump water the opposite way, from the Arctic Ocean through the dam to the North Pacific. They would draw off the Arctic Ocean's top layer of cold water. (Cold water is on top there because it is less salty, and so less dense, than warmer water which enters from lower latitudes. Freezing tends to eliminate salt from the surface water.) Theoretically if this cold top layer were continuously skimmed off, the Arctic Ocean would eventually be ice-free and the land around it warmer. The largest area to benefit would be Siberia. Some Soviet engineers say it would take three to four years for the Arctic Ocean ice to disappear. Others say the plan would never work.

Besides the differences of learned opinion already noted, there is another problem of the unknown. That is, would the warmer polar climate resulting from an ice-free Arctic Ocean really help northern lands or would the increased moisture in the air lead to more precipitation in the form of snow on surrounding land, then permanent snowfields, and ultimately regeneration of glaciers and ice sheets? If that process began these lands would be worse off than now, and who could halt or reverse the process? How could the open Arctic Ocean be made to freeze again? Some climatologists believe the ice cover exists on the Arctic Ocean only because it is there, reflecting away the sunlight, and that if it were once removed it would never re-form.

If Arctic regions were warmed there could be other side effects. Permafrost would thaw, causing much ground to sink as the ice in it melted, and farther south the climate might deteriorate. A warming of the North might be accompanied by a northward spread of deserts.

Another plan devised by USSR engineers is the damming of a strait between the warmer Sea of Japan and the colder Sea of Okhotsk to modify the climate of the Siberian coast across from Japan. They would also aim to improve the climate and agricultural potential of the dry and cold regions of their country by creating more new lakes and by altering river courses. They have already done some of this.

Since the main rivers of Siberia flow north to the Arctic Ocean, there is an excess of water in the cool, ill-drained North, but water is badly needed in southern, interior USSR, which is dry steppe and desert. The government is considering reversing some of these rivers. This would not be difficult because they flow over nearly flat, marshy plains. An inland sea may be created in Siberia to moderate the climate and add moisture to air masses that pass over it.

The United States is not without its own program to alter drainage systems. One is the Rampart Dam project devised for Alaska's Yukon River. A dam would be constructed on the Yukon northwest of Fairbanks and behind it a lake would collect. In less than twenty years this lake would be the world's largest artificial lake, larger than Lake Erie. Put a lake that size astride the Arctic Circle in east-central Alaska and the climatic reactions are bound to be significant. So far, this plan has been blocked because of possible environment-upsetting effects.

Australia is thinking of enlarging shallow, salty Lake Eyre in the desert by drawing in sea water to create an inland sea in the hope of increasing precipitation in adjacent areas and in the Australian Alps to the east. Lake Eyre was much larger during moister glacial times.

Some climate-modification plans tax the imagination. For example, Russian scientists are writing about the possibility of altering the solar radiation balance by creating a Saturn-like ring of dust particles several hundred miles above the earth. If climate modification became possible those with the know-how could direct climate changes to some areas and not to others. "Climatic warfare" could be waged. By creating floods or droughts, by increasing or decreasing cloud cover, one country might ruin the crops of another or upset military and civilian activities. Prolonged rains could start epidemics. Lingering high-altitude, artificially induced clouds could be laid over a particular region to diminish sunlight and be hardly noticed or be thought a natural condition. Someday huge plastic reflectors may be orbited in space above an enemy's land and be made to focus the sun's energy on a given location with killing heat. Miniature artificial nuclear suns might be hung in orbit to light the sky twenty-four hours and to warm cold regions. It has been suggested that space mirrors be used to melt ice caps and to light polar regions in their long sunless winters.

Melting of the ice caps would, of course, raise sea level. While this would be a catastrophe for most countries of the world with vital transportation and commercial hubs at ocean ports, the USSR of all world powers would hurt least. It has few good ocean ports and little good coastal land. Its only broad coastal plain is largely useless wasteland along the Arctic Ocean, frozen most of the year and otherwise marshy tundra, while coastal plains in warmer regions are some of the most productive and densely populated regions of the world.

There has been speculative talk of lowering or opening mountains, or deepening passes, to lessen the rainshadow effect caused by landform barriers—in Outer Mongolia, for example; and of gouging gaps in the mountain rim around Los Angeles to permit better air circulation and smog dispersal. Also of warming Hudson Bay with atomically energized heaters so it will not freeze. If the bay were open water in winter it would take some of the bite out of cold air masses before they reached southern Canada and the United States. It would also increase their moisture content and snowmaking potential.

Still more fantastic ways of changing climate, not yet in the range of possibility, are being thought of: tipping Earth's axis to another angle, changing the shape of Earth's orbit, placing Earth nearer to or farther from the sun, and slowing or reversing its rotation.

However, even simpler measures might alter climate. Air pollution, for instance. As mentioned previously, some atmospheric scientists believe air pollution will cool the climate, and some say it will warm it. Either could be right, for different kinds of pollutants do different things. In many areas, and perhaps the world as a whole, warming and cooling pollutants are both accumulating in the atmosphere at the same time, and since quantitative measuring of them has only lately begun, scientists have a hard time figuring out what the climatic effect is.

The National Aeronautics and Space Administration in a report presented to Congress warned:

The amount of carbon dioxide in the atmosphere has increased eight per cent in the last 60 or 70 years. Over this period there has been a great growth in industrial activity and in the use of the internal combustion engine. Since carbon dioxide in the atmosphere absorbs heat radiated from the ground, increasing carbon dioxide content implies a gradually increasing temperature at the Earth's surface. It would take only a few degrees rise in the average temperature of the atmosphere to cause profound changes in climate, the melting of the polar ice caps, with sufficient changes in sea level to inundate low-lying land masses such as Florida.

Air contamination from jet planes is a growing concern. As cars have helped smog up the lower atmosphere, so planes are polluting high altitudes. You have seen the smoky exhaust pouring out behind jet planes when they take off and the contrails they leave in the sky. These trails are, in effect, clouds of ice crystals, like natural cirrus clouds, and when many trails are spread close together across the sky they affect the incoming and outgoing radiation. Jets discharge less-visible pollutants also. Sky littering will become even more significant if supersonic planes come to be used, because in the high altitudes of the stratosphere where they would operate there is little vertical movement and the exhaust haze would stay suspended for a long time. Rockets will eventually be adding to high-altitude pollution too.

Temperature changes in turn affect wind patterns and precipitation distribution.

Now that it is known that the atmosphere can be manipulated, it will be manipulated.

We all acknowledge that short-term daily weather and long-term climate could be improved, and yet we feel a pang of regret as we foresee the passing of an era during which people could do no more than accept the whims and tirades of the elements. It was with some surprise that I found an expression of this nostalgia in the formal, businesslike report of the Special Commission on Weather Modification, National Science Foundation. Some member of the Commission felt a compulsion to tuck this sentimental remark among the paragraphs of objective, office prose. He wrote:

In the driving power of a winter blizzard or the sudden flash of summer lightning there are dramatic reminders of the elemental forces with which the human race constantly is striving to find its place. . . . A beginning at changing storm or lightning nevertheless raises the question of how far the human spirit is enriched by the uncertainty and wonder and exhilaration that come with the restless, violent movements of the atmosphere. Any effort to assess the social consequences of weather and climate modification must give weight to the esthetic and spiritual as well as purely material rewards.

It was strange to read that there, but it proves a scientist's outlook is not completely cold. "The driving power of a winter blizzard . . . the uncertainty and wonder and exhilaration . . . the need for the esthetic and spiritual as well as material." The challenge of tangling with nature's forces, which has been with us throughout the Ice Age and to the present, will be missed if ever we are "blessed" with living protected from their inspiring and terrible grandeur. But there will be a lot of bungling before the human race has the elements even slightly in control, and you can bet it will take a beating in the process.

Yet, people must experiment—as they have ever done—if they are to learn and develop their latent abilities and make their planet an even happier, more comfortable home.

But glaciers, so closely regulated by climate—what will be their fate?

CHAPTER 19

WARNINGS OF A WARMING WORLD

No attribute of man exceeds in importance his capacity to alter his environment.

IAN MCTAGGART COWAN

IN THE pages already turned we have looked at the Ice Age from many standpoints, esthetic and objective. We have seen what our Ice Age legacy is in material things of the earth and why we should preserve it, and how our human prehistory grew out of the Pleistocene past. We have looked at the older world before the ice came, and watched how the glaciers and ice sheets grew and withdrew several times, changing the face of the globe. We have imagined how our environment would be affected if the climatic pendulum ticked back toward the glacial period or tocked forward to an interglacial one.

With this broad understanding of the Ice Age and an awareness of our glacial heritage we are able to judge and appreciate what the world would lose if vestiges of the Ice Age were to disappear. The future of these vestiges—the ice's physical work on the landscape, the flora and fauna of that receding epoch, and the remaining glaciers—depends both upon the direction climate takes and upon our custody of them.

We shall look presently at ways in which human actions threaten our store of Ice Age inheritances, but first we should pursue our consideration of climatic change a little further to see how warming may come about. If it occurs it could be due to (1) a natural cause, or (2) inadvertent human interference, or (3) intentional human interference. It could well be due to all three in combination.

Consider first the natural trend of climate. We cannot look forward into the climatic future with any certainty but we can look back. It is as though we were sitting backwards on a moving vehicle. We cannot see what turn is coming next, but even though our course has been erratic we can see where we have been and thereby tell the general direction we are traveling.

Ever since the end of the Ice Age our climate has exhibited an over-all warming trend. True, there were setbacks to colder climate. The ice sheets had short readvances even during their retreat. There was the Little Ice Age and other cold periods, but we do not know that any of them were worldwide, and all were temporary. It seems plausible that this long warming trend should continue, heading toward the even warmer climate that seems to have been a more "average" condition. If it does, either the Ice Age has ended, or we are living in the early part of an interglacial period. (Some interglacial climate is known to have been warmer than the present climate.) In either case, we could expect that there would be a further temperature rise in middle and high latitudes with a continuation of the poleward migration of warm-climate vegetation and animals, including humans. There undoubtedly would be future temporary lapses to colder weather—perhaps a series of colder-than-usual years or even an occasional Little Ice Age—just as in spring there are recurring cold spells as the temperature climbs unevenly toward summer. But the long-term view widely held by glaciologists and geographers is that the warming trend will continue its wavering but persistent progression.

It is possible that this warming trend could reverse in the near future, of course, but it is just as possible that warming may accelerate. Climate has been relatively steady over rememberable time, so a jog in the temperature graph may be due.

Leapfrogging much farther into the future, beyond the anticipated period of increased warmth, we can envision another glacial stage because the ice has returned so many times, but before that happens, if it does, all or most of Earth's glacial ice may have melted away.

Speakers and writers seldom discuss the warming theory because it is not startling. What is startling is the opposite theory, that the Ice Age is about to return or at least that colder weather is descending upon us. Neither side possesses conclusive data to support its case, but those who preach that cold is returning have a much more eager audience. As they speak about temperatures falling and snows thickening, women in the audience shudder with excitement and men make worried jokes to each other and later discuss how their businesses will be affected. There is no scare, no apparent need to allocate funds or hire experts to prepare for or study a warming environment, and people do not sit on the edge of their chairs listening to what would happen in such a situation. Perhaps they should.

In his book *The Invisible Pyramid* Loren Eiseley relates a conversation he had, which I have used here to illustrate how strongly a person, a highly educated person, can espouse the idea that the Ice Age is returning:

"We have got to spend everything we have, if necessary, to get off this planet," one . . . representative of the aero-space industry remarked to me recently.

"Why?" I asked, not averse to flight, but a little bewildered by his seeming desperation.

"Because," he insisted, his face turning red as though from some deep inner personal struggle, "because"—then he flung at me what I suspect he thought my kind of science would take seriously—"because of the ice— the ice is coming back, that's why."

It is interesting to note that many of the theories that point to a return of the Ice Age use a postulated sequence of events that paradoxically begins with a period of increased warmth and ice melting, just what we are anticipating here. They would first have the Arctic Ocean become ice-free, or the Greenland and Antarctic ice sheets melt and the oceans rise and overflow their basins. This theoretically would make available a greater supply of moisture which with the prerequired higher temperatures would in a roundabout way lead to increased precipitation, particularly more snowfall in areas that could regenerate ice sheets. In other words, they would first get rid of all or much of the present ice in order to re-establish it. Meanwhile we or our descendants would be experiencing the warming, not the cooling. The new Ice Age would be the aftermath, but first would come the period of warming. There are, of course, countless versions of this theory.

Glaciers in general have been in retreat since the Ice Age, but those of Antarctica and Greenland are not good indicators of climatic change because they are surrounded by water and therefore never expanded to their largest possible size. We do not know how their present size compares to what it might have been in the Ice Age if they had been able to spread out on land. Their mass budgets are still unknown; that is, we do not know how much they are growing or diminishing. But in many smaller ice caps and valley glaciers recession has been obvious as they have shrunk back from their moraines. There have been readvances but these are outweighed by the retreats. We see mountain valleys that used to brim full of ice to the highest ridges now empty with only the scratches and drift and boulders left to show how deep the ice once was. We see mountaintops that were scraped by ice now standing bare. We see steep-sided fiords filled only with water and air. History describes settings that have changed with the glaciers' disappearance even within fairly recent times.

One such place is Glacier Bay in southern Alaska west of Juneau. In the 1700's that long, branching, mountain-bordered inlet off Icy Strait was completely covered with an ice cap 3,000 feet thick. No bay was visible. A cliff of ice stood along the coast. When Captain George Vancouver sailed by exploring that coast in 1794 he saw no Glacier Bay, just the seal of ice from mountain to mountain, and his map is drawn as though all behind were mountainous, ice-covered land. After his visit the ice slowly thinned and receded up the drowned valley. Many bays were reappearing in the same way but this one was especially impressive because of the many beautiful glaciers that draped its sides and the mass of ice that still blocked the head of the bay. Local Indians knew this bay, guarded by silver fogs, fleets of blue and white icebergs, and empty, stormy waters, but only rumors of its raw, primitive grandeur reached outsiders. Inevitably the stories reached the daring glacier enthusiast John Muir, and in October 1880, though winter was closing in, he urged the Indians to take him there. In a canoe they paddled through storms and icebergs to view the isolated, glacier-rimmed bay. Since Vancouver's exploratory voyage the ice had retreated fifty miles up the inlet. Muir found the panorama breath-taking as the canoe silently slipped into this bay that did not exist on maps. It was swathed in mist and on its walls hung an art-gallery array of valley glaciers, and he wrote how "lofty blue cliffs, looming through the draggled skirt of clouds, gave a tremendous impression of savage power." Since Muir's visit the ice has retreated even more. Sometimes the tidewater glaciers push ahead but more often when icebergs calve off, the glaciers fail to move up to reoccupy the vacated space, and more and more of the bay is uncovered. Now the great Muir Glacier, named for the famous naturalist and founder of the National Park system, has drawn back so far that it is dividing into smaller separated tributary glaciers. Fewer valley glaciers whiten the sides of the bay, and some lie motionless, black and dying on the terraces and mountainsides. But it is still a thrilling place because of its solitude and wildness, and because one can be in a true glacial environment there.

Polar specialist Bernt Balchen, who has been on expeditions with Amundsen and Byrd and has been active in the development of the Arctic, says that the Arctic Ocean may be ice-free by the year 2000, that ice over that ocean decreased in thickness from 43 feet in 1893 to a current average thickness of 6 to 8 feet in both winter and summer over an area of 5 million square miles.

Effects of warming can be seen closer than remote icy bays or the Arctic Ocean. Pieced-together scraps of statistics are often tendered as "proof" of a climatic oscillation, but real events which glide before us unnoticed can sometimes be more significant. There are heat waves and drought. And there are the Winter Olympics.

These international contests are held every four years, and the sites chosen for them are famous winter-sports areas that have an established reputation for heavy dependable snows preserved by low temperatures, because the snow and ice conditions have to be excellent and reliable. But what has happened in these ideal winter wonderlands in some of the last meets?

You may recall the 1968 Winter Olympics at Grenoble in the French Alps. Before the opening there was alarm because not enough snow had fallen to conduct the ski events. Snow finally arrived, but then just a few days before the start of events a warm drizzle brought a new scare. A February thaw forced postponement of the men's downhill ski races and bobsledding, and threatened to wash out tobogganing completely. The luge (sled) races require constant temperatures well below freezing, and the luge team sat waiting over a week, watching their course grow soggy. *Time* said it right: "One of the biggest problems at the Winter Olympics is winter."

Go back to the previous Winter Olympics of 1964 at Innsbruck, Austria. Opening day was January 29, and Europe was in the midst of a snow drought. The winter before had been that Terrible Winter,

Europe's worst in over a hundred years, when people swore the Ice Age was coming back. Now just a year later resorts were suffering tremendous losses as their slopes were green instead of white. Even Innsbruck was begging for snow. No snow fell in seven weeks preceding the Olympics. In balmy weather just before the opening 3,000 Austrian soldiers, Olympic officials, and volunteers hauled in trucks of snow from other parts of the Alps and packed the slopes. The games went on but the schedule was juggled in an attempt to get all events in. Many had to be held at dawn or before in hours of maximum cold. The bobsled and toboggan runs were slushy. Ski trails and a jumping hill were thinly handpacked with snow which barely covered the grass. Speed skating had to be delayed. Cars sank in mud.

The 1960 Olympics was held in Squaw Valley, California, selected because its average snowfall depth was 37 feet. But two weeks before the games pouring rains stripped the mountains of snow. Additional snow came just in time.

Four years before in Cortina d'Ampezzo, Italy, the Winter Olympics met a jinx too. It had not snowed for two weeks before the games; then came a promising snowfall but a thaw melted it all. Finally the Italian army had to truck in snow. Even so, casualties among the contestants were unusually high due to the skimpy snowcover. In one downhill ski race 75 started and only 47 finished, and there were 8 casualties. Scores of contestants received hospital attention. There were several brain concussions, numerous fractures, and dozens of torn ligaments.

Back to 1952. Oslo, Norway. The Norwegian organizers were confident. They had statistics to show that skiing conditions would be perfect. But two weeks before the opening Oslo had no snow. The slopes were bare. Practice runs had to be held off until snow was trucked in and the Norwegian army put courses in fair condition. Still a slalom race had to be shortened because the lower half of the course had no snow.

That takes us back far enough in the history of Winter Olympics to show that even locations famed for their winters are no longer strongholds of snow. Conditions have been changing.

Not only the Olympics have been feeling the snow shortage. Commercial establishments and public sports areas also experience frequent snow droughts. Skating rinks and snowmobile courses are often closed in mid-winter, and many a winter carnival is called off because of thin ice or no snow. Even Alaska has canceled dogsled races. Many ski resorts could not be in business if it were not for snowmaking machines. Some slopes are now made of plastic fiber so skiing can go on even when there is no snow. Somehow a future with diminishing snowfall appears sickeningly uninviting, and already we pity those who may have to live in it. We see skiers' button pins and car-bumper stickers that read "Think Snow" and "Pray for Snow" and our thoughts go beyond skiing.

But even if we should not be in a natural warming trend now, we may be drawn into one that the human hand engineers, unknowingly or knowingly. We are being warned that pollutants in the atmosphere may have a warming effect upon climate. Also Earth's radiating surface is being altered as civilization marches on, opening up expanses of dark soil, removing insulating vegetation cover and substituting heat-absorbing cement and blacktop, and melting ice.

The greatest concentration of fumes, concrete, and asphalt, the most striking transformation from the natural environment, has been the cities. Cities are spoken of now as "heat islands" because they are warmer than the rural surroundings. These furnaces of added warmth dot the landscape and are mushrooming and merging. The United States is now considered an urban nation. Urban areas have been expanding and coalescing to such an extent that some megalopolises cover a whole county, or several counties, or parts of several states. Mappers show the main urbanized areas of the near future as agglomerated blotches hundreds of miles wide.

London, once the world's largest city, was recognized as a heat focus back in 1818. Now we realize that small towns too have this characteristic, though to a lesser degree. Even a group of farm buildings has some small effect. The temperature difference between a built-up area and the country is related to the size of the built-up area; and the intensity of the heat island increases as the city grows.

Seldom has there been a chance to measure the rural-to-urban change, but a good opportunity came with the building of the brand-new town of Columbia, Maryland, northeast of Washington, D.C. Within two years the site changed from country to city, and in 1970, when the population had reached 7,000, marked effects were already noticed. Wind velocity had decreased, pollutants had risen 10 per cent, and air temperatures had risen 2° F. above that of the country. Helmut E. Landsberg of the University of Maryland, who made de-

tailed meteorological studies there, views the results as consistent
with climatic changes resulting from urbanization elsewhere. He
says, "With accelerating urbanization one can predict that local climatic changes will increase from a local to a regional scale and will
appreciably affect the ecological balance."

In cities on a summer day solar radiation is absorbed by buildings,
streets, and parking lots with large heat-storage capacities. There is
little vegetation or moist surface where heat would be consumed by
evaporation. Cement, brick, stone, asphalt, and metal surfaces absorb and radiate heat; buildings block ventilating breezes; smog
makes a tent over all; and the city grows hotter than the country.
Comes evening, and the temperature begins dropping in the clearer
air of the country, but the city's buildings and streets cool slowly
and keep radiating their stored heat well into the night. The city has
a head start in temperature the next morning.

On a winter day heat seeps through buildings' walls and windows,
and pours from chimneys all over town and from factories, power
generators, cars, and buses. At night furnaces keep producing heat,
and unless there is a strong wind it is held there by the trap of buildings and smog. Even body heat is not insignificant. An electric company figured out that a person gives off as much heat as a 50-watt light
bulb, and that in one day 20,000 people exude 5 million BTU's (British thermal unit—the amount of heat needed to raise the temperature
of a pound of water one degree Fahrenheit).

A study in Vienna showed that heat from furnaces and electric
power there was about half that received from the winter sun. In
Manhattan in a recent winter the heat from combustion has been
found to be two and a half times the heat from solar radiation.

Meteorological data have revealed the surprising fact that the frost-
free growing season in central Chicago and in Washington, D.C.
(and so undoubtedly in many large cities) is now a whole month
longer than it is in the rural area outside the suburbs.

San Francisco, famous for its temperate climate and comfortably
cool summers, is hurrying the heating of its area. Its picturesque bay,
used as a dump, is being filled in little by little. A meteorologist has
predicted that filling 70 per cent of the bay will lessen the marine
effect enough to raise summer temperatures 10° F. Cooling breezes
will diminish and smog will increase as buildings are constructed
on the fill.

The influence of a city does not halt at the city limits. The smoke
pall of a large city affects heat-radiation measurements over an area
50 times that of the urban center. Large metropolises spread a down-
wind veil of smog as much as 500 miles over the countryside. W. O.
Roberts in the *Bulletin of the American Meteorological Society*
writes of flying east out of Chicago and observing a streak of gray
haze that emanated from the Chicago-Gary area. It reached almost
to Washington, D.C.

Hamlets, towns, and cities are spreading like prairie fires, warming
the environment in their limited way, and they are not all distant
from glaciers.

So climate may warm naturally. Or it may warm as mankind un-
intentionally changes the atmosphere's composition and transparency, alters the rural landscape, and builds expanding urban heat
islands. If warming fails to come about in those ways it may still be
brought about as mankind develops the ability to modify climate
by deliberate design. It is this last method of warming—deliberate
action by mankind—that may hold the greatest warning of a warming
world. Its effect could take hold more rapidly than that of natural or
inadvertent warming.

In small ways we already have numerous devices to warm corners
of our outdoor environment. Outdoor heaters make our patios comfortable on cool nights. Sidewalks and pavements are electrically
heated, and snow melters and flame throwers are used to get rid of
snow and ice. Birdbaths are warmed to keep their water unfrozen
in winter. Electric cables warm football fields for late-season games.
Some buildings provide outdoor heating at entrances and places
where people congregate. These warming techniques are Lilliputian
but they are a sign of our desire to fight snow and cold, and we do it
wherever we can.

Larger-scaled methods of changing weather and climate, which
are in the experimental or planning stage, were mentioned previously: cloud modification, storm control, redirecting ocean currents
and rivers, creating lakes, unfreezing oceans, cutting through mountains, using space mirrors, and so on. In addition, there is talk of
achieving warming by asphalting large areas, by floating long sheets
of plastic in the sky to produce a greenhouse effect, and by warming
oceans with atomic heating plants or detonating H-bombs under
water. Most of these ideas are still farfetched, of course.

People have fought to overcome every natural obstacle they could in their relentless conquest of the land of the world. Bitter cold with its ice and snow is the most formidable and unyielding obstacle they have had to contend with. Their war with Cold is only now beginning in earnest.

If the human race is not thrown backward by some catastrophe it will move ahead and learn the techniques of climate modification. When one has a good tool or skill he is eager to use it, and so as soon as these techniques are developed they will be applied, even though they may not be mastered. It takes no seer to predict that where temperature is concerned the first big item on the agenda will be the combating of cold.

As things are now, the countries that stand to benefit most from warmer climate are the northern ones of the Northern Hemisphere. The United States with Alaska just awakening. Canada with its vast Arctic wastes. The USSR with Siberia, the coldest region not covered with ice. Teeming China with its cold northern provinces bordering on Siberia not nearly as productive as they could be. Damp regions of northern Europe, another of the world's densely populated regions, whose agricultural production could improve with warmer weather and whose northernmost reindeer lands are too cold for general occupation.

And these regions with severe winters or limited growing seasons or resources frozen most of the year are world leaders today. They are the seats of power; they have the scientific experience and capability; their people are the ones who would probably be doing the climate or environment changing if it were possible. They would have the say-so as to which way temperature would be steered, and certainly they would not choose greater cold. They want "improvement," "progress," "easier living," and this would call for warming. With warmer climate they would have greater self-sufficiency in food, fibers, and minerals, more comfortable living space for their growing population, and a wider sphere of influence.

From the foregoing we may assume that if warming were to occur the scientifically, politically, and militarily dominant northern countries would make no attempt to stop it. However, if cooling occurred their governments, private enterprises, and scientists would work to reverse the trend if only to maintain the climatic *status quo.*

A case in point is the speedy cooperative action taken a few years ago by the not-always-compatible Americans and Russians when it appeared that the Arctic Ocean ice cover was thickening and becoming stronger near the coast. Both would like to see that ice thin and melt back from the coastlines now even more than previously because of the oil fields on Alaska's north coast and the valuable ore deposits in northern Siberia. Both countries may want to transport these commodities out by sea. They wasted no time in stepping up investigations to see why the ice thickened and what could be done about it. They have used every exploratory device available, including nuclear submarines, aircraft, space satellites, and numerous testing stations on drifting ice floes. More powerful icebreakers are being designed. And if the ice should continue to strengthen one can anticipate direct action being taken to combat it.

If alarmists repeatedly scare us about the climate cooling, even when they do not really know that it is, weather changers may be spurred to aid the warming processes. The signals may have been misread, we may be wrongly informed, and we may even be on the warm track already. Then intervention to counteract cold would just add impetus to an already existing warming trend.

We can be sure that if unmistakable, pronounced cooling does actually take place, the greatest forces of science will be thrown against it. But if gradual warming takes place it will probably be ignored indefinitely, even welcomed.

Even if all that we have been assuming about climatic warming never takes place and conditions stay much as they are now, our glaciers are still in danger. Individual ones and maybe all of them stand in eventual peril of being assaulted quite directly. If climate does not harm them mankind's physical actions may. To be attacked by both at once could be a glacial tragedy.

CHAPTER 20

THE FRESH-WATER CRISIS

Water soon will be the most valuable mineral resource we can mine out of the earth.

ATHELSTAN SPILHAUS

TO SAY that glaciers are in danger of being destroyed sounds extreme. Glaciers seem as ancient and eternal as mountains, but of course they are not. Those dwindling Ice Age holdovers are transitory accessories of this planet and exist for a while only because of a favorable temperature-moisture balance in the lands that support them.

When dense virgin forests and flowering prairies covered an area of North America as large as the Greenland ice sheet, who would have thought it likely they could ever be destroyed? Yet in a few generations they were rooted up and gone. All that was needed to remove them was a will and the physical means.

There already exists a will to move in on the glaciers, but it is seldom given open expression, for the idea sounds preposterous. The means to melt them are becoming available—nuclear power, climate modification, heat-focusing space mirrors, and dusting with black powder. The last method has been used for years.

The main human threats to glaciers stem from the population explosion which has led to a plague of hunger and poverty, a depleting of Earth's resources, and a deteriorating water supply. There are other threats besides, as we shall see later.

We are well educated to the fact that without birth control Earth's population will double about every thirty-five years. Even if birth control has some success, death control is even more successful; and

Earthlings, who call themselves the smartest of the animals, keep multiplying at an accelerating rate while the worst food famine in history looks them straight in the eye. Less publicity has been given to the impending water crisis which soon will be as alarming as the food shortage. We are aware of the afflictions of drought and pollution, but they are only part of the water-crisis story, a story in which glaciers play a leading role.

Water—fresh water—is the most valuable of our natural resources. It is becoming one of our scarcest. Though we call ourselves the Water Planet, most of our water is not the kind that can be used for personal and home use, for agriculture, industry, and maintenance of the environment we live in.

Over 97 per cent of the planet's water is salty ocean, unusable for those purposes. That leaves us less than 3 per cent. The atmosphere holds a small 0.001 per cent of the total. Only about 2.8 per cent of the world's water is on land, and some of that is saline lakes. Some is in the ground—0.6 per cent if we include that down as far as two and a half miles, and much of that is too mineralized for some uses. Fresh-water lakes hold a meager 0.009 per cent, rivers a wee 0.0001 per cent, and practically all of those in populated areas are now polluted. All that remains is in the usually disregarded glaciers which have 2.14 per cent of the world's water—sweet and clean. That figure stands out like an oasis in the desert.

Glaciers hold the world's greatest reserve of fresh water! Over three fourths of the world's total supply! And it is pure!

These reserves have been left undisturbed because of their distance from heavily populated areas and because water demands could be met from closer souces. But the population is spreading closer to glaciers and water needs are spurting upward. Water's cost is rising due to the growing demand, the increased processing required to depollute it, and the need to draw it from ever greater distances. There are substitutes for many commodities but not for fresh water, and it is basic to life—human, plant, and animal.

Water is more critical to life than food. One can survive weeks without food but only a few days without water. Just to survive the average person must consume about two quarts of water every day as a liquid or in food. If we consider all water requirements of our society—municipal, industrial, agricultural, and so on—the need in the United States is over 1,500 gallons per person per day, and our rate of use is increasing more than twice as fast as the population is. The amount of water used in countries around the world is also sure to mount. In some developing countries up to 90 per cent of the population now have inadequate piped water service or are being supplied with unsafe water. Many people cannot even buy the minimum amount of water needed for normal living or they obtain water directly from wells, rivers, and other sources open to contamination. As these areas develop, their demand for water for all uses will soar.

In years ahead the present developed countries—as well as the better-developed regions within a country—will not be the only gluttonous gulpers of water. Their less-developed neighbor countries and communities will be calling for more water too for stepped-up industrialization and agricultural production, and for improved fire protection, sewage disposal, better-kept yards, more recreational facilities, more air conditioning, and so on. Some families may be aiming just to have running water in the home while others will be aiming for a private swimming pool, but everywhere water consumption is due to increase enormously.

The populations of underdeveloped areas are multiplying three times faster than food production in spite of foreign aid and improving agricultural techniques, so malnutrition and starvation are not only unchecked but on the rise. New land not now cultivated or pastured will have to be brought into production, and that already producing will have to step up to a still more intensive type of agriculture—higher yields per acre, more crops per year, closer plantings, vegetable farming instead of grain, and so on. This will require irrigation. Most of the best arable land in humid areas is already farmed. The always-cold lands cannot be. Dry lands hold the greatest promise for supplying additional crops, animal products, and fiber. A migration has already begun into those empty regions and not only by poor people. And not only in backward, overpopulated areas, for population pressure is increasing everywhere. Dry lands have great potential. They just need water.

One third of the world's land is in the arid realm—grasslands and desert. Arid lands have some advantages over humid ones for plants and animals. They are spared many insects and diseases and have more sunshine. Ripening and harvesting of crops is not hampered as much by untimely rains and dampness. Grassland soils are exceptionally fertile—the most fertile of any residual soils. But irrigation requires a large, steady supply of water, much of which evaporates and soaks unused into the ground.

Mountain glaciers are an important source of irrigation water. We have seen that they are commonly located along the edges of dry lands, an increasingly vulnerable position. The high, glaciered mountain complex of southern and eastern Asia supplies irrigation water to dry intermontane basins and surrounding lowlands from Asia Minor through Pakistan, India, and southern USSR to China. Plains, plateaus, basins, and valleys of western Canada and northwestern United States use irrigation water from rivers that head in glaciers. The Andes, Alps, Caucasus, and other glaciered mountains also supply needed irrigation water.

Irrigation is practised in humid lands as well as arid lands. Anyone who has sprinkled a dry lawn or trickled water around a wilting garden plant has resorted to irrigation and knows that just a little extra water at a critical time can do a lot of good. Many rivers flowing through humid lands are fed in part by tributaries heading in glaciers.

In a previous chapter we have seen how governments, farmers, power companies, or other experimenters in a number of countries have been dusting mountain glaciers with soil, ash, or other dark substances to force them to melt faster. Since mountain glaciers are already being artificially melted one may peer ahead to a time of denser populations with more insatiable water appetites and imagine how even such faraway "mountaintops" as Antarctica or Greenland might be called upon to slake the world's thirst. Fantastic? Considering their inaccessibility and the problems of transporting water, perhaps so—for now. But the inaccessibility and transportation problems are diminishing with the advance of science.

Look at the quantity of water these glaciers store. The U. S. Geological Survey gives these figures: The Greenland ice cap, with its volume of 630,000 cubic miles, if melted could yield enough water to maintain the Mississippi River for over 4,700 years. The water in the Antarctic ice sheet, which contains 85 per cent of all existing ice, could feed the Mississippi at a uniform rate for over 50,000 years, or all rivers of the United States for about 18,000 years, or all rivers in the world for 830 years! The lesser glaciers seem insignificant beside these giants but they are not so, for the rest of the glaciers and ice caps hold as much water as that in all the lakes—fresh and salty —in the world. And their virgin water is purer than any lake's.

Planners touch only fleetingly upon the subject of artificially melting glaciers for future water supplies, but we get the disturbing feeling that our glaciers will be made to deliver the goods even as mines and forests and hunted animals have had to do until they were used up or extinct. People who suggest using them do not have any such tragedy in mind. They are thinking of the human good. But they are pushing glaciers from their untouchable status to the unfortunate status of a usable resource.

A U. S. Geological Survey spokesman is of this opinion:

Glaciers and perennial snowfields are, in effect, nonstructural water reservoirs, and the possibilities for their management merit thorough investigation. Possibilities include suppression of evaporation, suppression of melting in wet years, *inducement of melting in dry years,* and others.*

The forward-looking scientist Athelstan Spilhaus wrote in the *Geographical Review* back in 1956:

As pressures arise to inhabit the more arid regions of the world, means of bringing plentiful fresh water to these regions must and will be found. Imaginative engineering must discover economical ways to supplement natural distillation from the sea and *find means of using the fresh-water ice that nature provides in abundance in polar latitudes.*

Science writer D. S. Halacy, Jr., wrote in his book *The Weather Changers:*

As population overflows the regions naturally favored for habitation, it would be wonderful if we could make the deserts "bloom as the rose" and *the frozen north melt to springtime warmth.*

President Lyndon B. Johnson said:

The world toward which we are all working is one in which the earth will yield up enough for every man in every country. A world of peace in which the very deserts bloom and *the polar ice is turned to enrichment of man's life.*

* The italics in this and following quotations are mine.

In the National Science Foundation's report of the Special Commission on Weather Modification, *Weather and Climate Modification*, it was stated:

Should the need grow so desperate or economical means of transportation be devised, *ice and its melt water provide an as yet unexploited resource.*

Glaciers are frozen reservoirs. In an emergency our policy is, of necessity, to let reservoirs go down, empty if need be, and then refill them later. But glaciers may never fill up again.

For the time being there are still conventional methods of meeting the growing water demands.

Rivers are dammed to impound water and keep it from running off the land before it can be used. But these lake reservoirs flood large areas, often of good farmland, and evaporation consumes much of the stored water. These dammed lakes begin silting up as soon as water starts collecting, for when inflowing streams lose velocity they drop the silts and sands they are carrying. So new dams continually have to be built, and they are being made bigger all the time. Glaciers—those immense natural reservoirs—need no cement walls to hold them, no maintenance, and do not silt up.

The ground is another source of water. Wells run dry, so we drill deeper or drill somewhere else. Artesian wells, which flow continuously under their own pressure, give an illusion of endless rivers underground. Stories of discovered "lakes" of underground water—saturated porous rock—are also misleading, for these reserves are not all renewable. Over most of the earth rainfall can be relied upon to replenish the groundwater supply about as fast as it is used, but there are areas that are not being recharged. Some of the largest underground water reserves—particularly in arid regions—contain water that has collected over thousands of years. Those aquifers are now yielding "fossil" water that fell as rain during moister periods, much of it during the pluvial periods of the Ice Age. Use it, and it will not be replenished. And it is being used.

Some groundwater has an objectionable taste, smell, or color. It may be "hard" or highly mineralized, undesirable for many purposes. Where too much water is pumped from the ground in places near the sea, salt water seeps in to replace it and then the wells turn salty and useless. If water charged with salts is used to irrigate land it will quickly ruin the soil.

Glacier water is soft and sweet as rain water.

Desalination of sea water is a great hope, but it has its drawbacks. It is still an expensive, inefficient process, so it is resorted to only where water is critically short. There are a number of different converting processes, and not all produce water of desired quality. Wherever the desalted water is distributed it has to be pumped up, since the oceans are the lowest level, and the farther it has to be transported from the coast the more its cost increases.

Mountain glaciers are in the highest places and distribution of their water for considerable distances requires no power but gravity. No expensive conversion either. Simply melt and use. And the melting process could be as simple as dusting by airplane.

Re-use of water is a coming way of fighting the shortage. Much used water will be channeled back, cleaned up, and used again. But every water use—agricultural, domestic, and industrial—loads the water with more chemicals. Some water is so dirty and contaminated that the high cost of purifying it would prohibit its re-use. And people do not like the idea of drinking such reclaimed water or using it for domestic purposes. Viruses are extremely difficult to get rid of, and as water is recycled its concentration of viruses is likely to become greater.

After being obliged to use such unappetizing water, how tempting good, clean, fresh ice water would seem. It might be worth moving one's family—or industry or farming operation—nearer to glaciers (the closer ones of temperate climates, of course) to obtain such water. Some industries and processing plants would find advantages in using glacial meltwater because of its purity and coolness. Over 90 per cent of industrial water is used for cooling. Many establishments are moving to outlying locations formerly thought illogical to escape high taxes, labor difficulties, problem-packed cities, or to obtain cheap land to spread out on. In wartime decentralization is encouraged for security.

Glacier water has no troublesome or offensive impurities. It is soft and practically sterile. Ice layers near the glacier's surface contain a little fallout from air pollution, but this was nonexistent during most of the time glaciers were forming, before the industrial and scientific revolutions, so the inner, unbroken layers are uncontaminated. The larger, more remote glaciers are exceptionally pure. Analyses made of water taken from ice in Greenland and Antarctica showed it to be nearly comparable to the purest laboratory water. A bacterial count at the South Pole showed only one bacterium per pint of snow.

Glacier meltwater contains ground-up rock and earth materials, but so does every river and lake, and this is "clean dirt." Settling eliminates it. Most of these materials have been dragged at the ice's base. Inner layers contain little if any foreign material. Meltwater flowing from a glacier's base may be muddy or turbid if the glacier is active; but slow or stagnant glaciers discharge relatively clear water. The heavier material drops out quickly in the outwash zone. Glacial lakes are often crystal clear. Some glacial streams have a milky look because they contain finely ground rock flour. This powder is the last material to drop out of the discharged water and, while suspended, it gives a beautiful turquoise color to some glacial lakes.

Now, it may seem that by advertising the superior qualities of glacier water I encourage glacier exploitation, but these facts are already known by the main water users, regional planners, and agriculturalists who have access to glacier water. In years to come we shall appreciate glaciers more than we do now. We should anticipate that pressure will be put on planners and governments to allow over-use of glacier water, make it serve larger areas than it naturally would, and spoil many glacier settings in the process. The danger is that because the "interest," the meltwater, from those securities improves living conditions as much as it does, there will be a growing desire to draw on the "principal" too.

We delude ourselves if we think glaciers will be respected by water users. Hardly any river or lake has been. Distance and physical obstacles are no longer shields. When water becomes dear enough so that consumers will pay the price to get it from far away, the technical problems of transportation will be mastered.

We are not thinking in terms of sending water from Antarctica to the Sahara, or from Greenland to the Gobi, and we do not imply that glacier water can be sent anywhere it is needed. Little local glaciers are perhaps more important to their areas than Greenland is to the world. They may need protection in coming years if they are to last.

One could say there is no actual worldwide shortage of water, that the problem is just one of poor distribution. But what a problem! Too much of the good water is where people are not or where the need is small. Water-deficient areas are drawing water from the handiest surplus areas, but as population spreads to those areas the surplus will no longer exist and water will have to come from still farther away.

The concept of transporting water great distances is not hard to grasp when we think of what has already been done. Way back in Roman times water was brought by aqueducts to Rome from the glaciered Alps hundreds of miles away. Thousand-mile pipelines transport oil and gas cross-continent, and these commodities are not nearly as vital as water. Water is already being routed long distances to populated areas by shifting it from one river system to another, by using aqueducts and canals, by pumping it over divides or sending it through mountain tunnels.

The Columbia Basin Irrigation Project was an engineering marvel when first executed. It uses the Columbia River, which heads in Canada's icefields, to power Grand Coulee Dam and irrigate Washington's dry plateau. Now more ambitious plans are being made for that area.

Arid parts of California with a precarious water situation have had to reach for water to more humid parts of the state and to the Colorado River, but those sources will not long be sufficient. California has millions of acres in irrigation. The influx of people to that state has made it the most populous state in the Union. Her population has been increasing about 1,500 a day. All of the Colorado River's water is already apportioned to other western states, so California cannot hope for more water from it. Desalting of ocean water is expected to supply only a fraction of what is required in the years ahead. So California is coveting the next closest source, the bountiful Columbia River. It flows south from British Columbia and forms part of the boundary between Washington and Oregon. That Northwest area has an oversupply of water at present but its people do not want to divert any south because if California continues increasing in population at that rapid rate she will quickly increase her dependence upon their water and need to continually enlarge her take. As more water is supplied to California more people will be encouraged to come there, and once they are there they must be cared for.

The Los Angeles *Herald Examiner* said:

What's good for the goose is good for the gander, and if you are going to take water away from Colorado why not take it away from Washington and Oregon when the taking of normally wasted water is vitally necessary for the existence of arid areas?

California has hopes of tapping the Columbia River one way or another. One plan is to run a plastic intake pipeline along the Pacific shore, in the ocean, up to the mouth of the Columbia where it will catch the discharge before it enters the ocean. How much farther north might such a pipeline be run someday?

California's former senator Thomas H. Kuchel had this to say:

As an increasingly thirsty Pacific Southwest becomes a vast megalopolitan complex, it is compelled to look, perhaps afar, for a new water supply to slake its thirst. That water supply could be as far away as the Klondike.

The glaciers begin to seem not so remote.

The high plains of Texas and New Mexico are also urgently seeking more water. They are eyeing the Lower Mississippi River 1,000 miles away, even though the water would have to be lifted about 4,000 feet and cross rough plateaus, and they are also looking at the possibility of acquiring water from Canada.

One reason arid agricultural lands need an ever-increasing copious supply of fresh water is to forestall the build-up of salts in the soil. Irrigation adds salts to the soil, and the eventual salt concentration becomes so great that the soil is unusable. As water flows across arid land it picks up salt which accumulates where evaporation exceeds precipitation. Each time a river's water is diverted to a field and drains back into the river it acquires more salt. Because of this, and evaporation, a river's saltiness in arid regions increases as it flows downstream.

Early civilizations that relied on irrigation collapsed ultimately because of salt build-up in their soils. It happened in the Tigris-Euphrates Valley of Iraq and in many other areas of the Middle East. Salt accumulation is already afflicting newly farmed lands irrigated by the Aswan Dam in Egypt. It has ruined much land in India and in irrigated regions elsewhere. The Rio Grande Valley, southern

California, Arizona, and other parts of the West where irrigation is practised have not escaped this serious problem.

The Imperial Valley of Lower California, a principal supplier of fresh fruit and winter vegetables, can produce up to five crops a year through irrigation but it is dying by that very irrigation, poisoned by excessive salts. It obtains its water from the Lower Colorado River near the Mexican border after the river has picked up salt from seven dry western states. Carl Bevins, president of the Imperial Valley Irrigation District, reported that thousands of acres have already been withdrawn from production because of the cost of fighting salinity. He said:

Somehow, from somewhere, we've got to get better water. It could come from aqueducts, transporting water from the big rivers in the Northwest or Alaska.

Chambers of commerce, real estate developers, and local governments invite families and industries to settle in some water-short areas (not only in this country and not only in arid regions) even when they do not know where future water supplies will come from. There exists a mystical assurance that there will always be enough water.

Major General Jackson Graham, formerly with the Corps of Engineers, while working with the Bureau of Reclamation asked, "Must we always try to bring water to people, no matter where and in what inhospitable regions they may choose to wander? Is a man entitled to buy up, settle, or promote a chunk of desert and then demand that his government bring water to him from the general direction of the North Pole?"

After people have settled in an area, invested in homes, land, and businesses, and developed emotional attachments, they cannot just be sent away if water problems arise, nor can they be expected to leave voluntarily. When the deepest wells run dry, when neighbor communities can no longer increase the water allowance, when rain- and snow-making does not work, water must come from somewhere. If your personal life were affected you might be one of those insisting on action. When your water begins to smell and taste funny or foam at the faucet, when it costs too much and water pressure is low, when you are forbidden to sprinkle your lawn, or fill your pool, or

wash your car, then you might not think it such a bad idea to dust
that glacier up there so that it melts faster, or to hire a weather
manipulator to try to make rain fall on your land even if that means
glaciers in the mountains beyond might not receive their quota of
snow.

Sometime, assuming the population keeps growing and the empty
lands of the world keep being settled, water emergencies will become
so acute that many of our glaciers will be wrung out. We should make
a realistic appraisal beforehand—there is still time for that—and decide how much of our glacial inheritance we are prepared to sacrifice
for poor planning. We cannot go into the problem here of whether
to, or how to, try to "manage" glaciers the way many renewable resources are managed so they can be used and replenished, but this
matter will have to be faced in the future.

We can count on some glaciers being lost, used up, because of lack
of foresight and unplanned-for emergencies. The situation may be
such that the natural melt of glaciers will be sufficient in a given area
in normal times, but then comes the unexpected. War—when a country must be self-sufficient, when more power is required, when desalting plants or aqueducts may be bombed. This is a matter of
survival. Or drought comes, with all its attendant inconveniences
and hazards. Irrigation pipes drip dry, health is endangered, work is
slowed, and then housewives picket and special-interest groups pressure for more water from anywhere. Or salty fields and polluted lakes
and rivers need to be flushed out to be saved. Or the population has
taken an unexpected upswing.

We would not think of pumping wells dry, but we have. Or destroying our lakes, but we have. Or using up glaciers, but—that time
may come too.

In our search for water we go ever closer to the mountains and ice
caps whether they be a hundred miles away or a thousand. It is not
that we are after the glaciers particularly, but where they are the
streams flow when others do not. People cannot help wishing at
times that the glaciers would release more water than they do, and
they may go after it. "They" could be ranchers in America, peasants
in China, plantation owners in South America, or power companies
in Europe, or someone else.

When glaciers are on mountains along borders, as they commonly
are, the people on both sides need them. In times of ill-feeling

or war or other crises no friendly agreement guarantees their protection. No government in the world can be expected to be altruistic, conservation-minded, and honorable at all times. Even less can be said of individuals or groups intent upon a greater income. A glacier or icefield could be dusted surreptitiously to increase streamflow, and there would be only slight graying, invisible to the eye. However, a plane flying over with remote-sensing devices that detect heat radiation differences could discover where dusting had been done.

Consider the glacierized mountain complex of High Asia. It amounts to over 50 per cent of all glacierized areas outside the polar regions and is 33 times larger than the Alps. Too big to be hurt probably except around the fringes. But on the south lies India, adding a million people every month, and other hungry peoples that rely on river water from those mountains. On the east, China. Her agricultural expansion in the arid west is based on glacier-fed rivers and she is already dusting and blasting glaciers to increase the water flowage. On the north the Soviet Union, intent upon developing her dry land in central Asia, but the steppes are often a Dust Bowl and the important Aral Sea is drying up. More water must be had there.

In Europe and Latin America too glaciers share their water with more than one country, and each may have a different regard for the glaciers on their mutual border.

Small, scattered glaciers of temperate regions are not large enough to be counted on for forced delivery of additional water very long if they are not replenished with snow. Even if they are called upon only in critical dry years this might be enough to cause their extinction. Or the cause of death might be something less intentional—a nearby forest removed; new smoky chimneys in the valley; or a large warm-water reservoir. Their disappearance would be of little concern probably. They would just gradually shrink and one spring they would dissolve away, as many are doing right now. The mountains could still be snow-capped most of the year and who would notice . . . except those glacier lovers who watch the mountains and look every summer when the snow vanishes to see if the familiar irregular shapes of white still remain spilling their silvery threaded waterfalls down the rock cliffs.

We are of the same breed as the "sportsman" who has chased the polar bear, an endangered species, over ice floes with helicopter and rifle till the bear is exhausted; who has hunted the mountain sheep and mountain goats to their last refuges in the stony high slopes, and the beautiful white, spotted snow leopard to its last lairs in the Himalayas. Will glaciers too be pursued to their last polar and mountain refuges?

Long-distance water-distribution schemes become broader and more imaginative all the time, and engineers have the techniques and power to implement them.

The proposed North American Water and Power Alliance is a colossal water-distribution plan that would pool the waters of the continent. It would take water flowing from the mountains of western Canada and Alaska, and through a system of river-linking canals, of hundreds of dams, of power plants, tunnels, and reservoirs, some as large as Lake Erie, it would recirculate the water to seven Canadian provinces, thirty-five states of the United States, and three states of Mexico! This would relieve the water situation of the West and of northern Mexico for about a hundred years—that is, temporarily. The network would pour more water into not only the Colorado and Columbia rivers but the Mississippi-Missouri and the Great Lakes too. A ship could travel across the continent from Vancouver, British Columbia, through the network of connected rivers and artificial lakes to Lake Superior and then through the Seaway to the Atlantic.

Canada is already laying groundwork for redirecting some of her northward-flowing river water which empties into the Arctic Ocean so that it will flow to her southern prairie provinces instead. But she is understandably reluctant to export her water beyond her borders, even for a good price. Her resistance, however, may be lessening; she already diverts water to the Great Lakes to maintain their level, for they are as important to her as to the United States.

A sign of the seriousness of America's water shortage is the appeal that came from the Office of Science and Technology in Washington a few years ago. It asked for "research on far-out ideas" to increase water resources.

One forthcoming suggestion, from Robert D. Gerard of Lamont Geological Observatory, was to dam the east and west ends of Long Island Sound to create a 150-mile-long fresh-water lake between the northern shore of Long Island and the shore of Connecticut and New York. By keeping out salt water and allowing only fresh water to enter—from rivers, precipitation, and groundwater—the water would become fresh in about seven and a half years, and the New York metropolitan area would have the United States' largest reservoir at its doorstep. One problem is that pollution would be pouring into this reservoir from the heavily populated and industrialized areas around it.

Worried Easterners keep looking to Canada for water, for she has a superabundance—if only it can be distributed where needed. Some Canadians are suggesting diking off James Bay, the southern protuberance of Hudson Bay, to turn it into a reservoir of fresh water brought in by inflowing rivers, and then channeling this water to Lake Huron to help stabilize the level of the Great Lakes and make more water available to both Americans and Canadians of that area.

So the searchers for water in middle North America look from the dry Southwest and Mexico up to the Columbia River, then the Canadian Rockies, then the Klondike and Alaska for more fresh water, and in the East to eastern Canada and Hudson Bay. The next jump is Greenland, but who would consider that?

A New York Times reporter came close during the last drought when he started his news story this way:

So far, no one has formally proposed easing New York City's continuing water woes by tracking down and towing into port a mountain-size North Atlantic iceberg—which might then be tapped for a time as a kind of floating water works.

But it may not be long.

He was not the only one whose mind has drifted in that direction.

The quest for water brings us ever closer to the strongest citadels of ice. We have crossed the ocean moats that protect them, we are overpowering the armadas of icebergs and floes that guard their gates, and are already chipping at their outer walls.

CHAPTER 21

ICEBERGS, ANYONE?

And the grand icebergs!—so cold, yet so majestic; so solid, yet so unsub-
stantial; so massive, yet so ethereal! . . . The human mind can not con-
template them without a sympathetic inspiration, for their duplex entity
is so like our combination of soul and body!

CHARLES HALLOCK

GLACIERS and their offspring, icebergs, have always
carried an aura of danger and have been avoided by all except the
most adventuresome. It is known that people and animals who stray
onto glaciers may fall into crevasses or slippery drainage holes and
never be seen again unless their frozen bodies are preserved in the
ice, carried along undisturbed, and redelivered at the melting edge
years later. Ships that venture too near icebergs may collide with
them in fog or even in clear weather, for about 85 per cent of a berg
is submerged and that part may extend out far beyond the visible
top. A ship that comes too close may capsize in churning water when
an iceberg suddenly rolls over as wave cutting and unequal melting
upset its equilibrium. A berg may turn over several times a day during rapid melting.

However, the fear of ice is abating. In spite of its dangers, which
are real, the ice is beginning to seem not inhospitable and repelling,
but inviting and refreshing. Today most people of the world live in
sweaty congestion, choking in miasmic and dusty air that careless
industry and agriculture have blown about them, shrinking in disgust
from the uncleanness which surrounds them in overcrowded, stagnant
living areas. Now the thought of floods of ice—untouched, crystal
crisp, glistening pure—in an invigorating setting where all is as un-

sullied as when it was created makes us crave to be in such a vitalizing place or, failing that, to bring that refreshment to us. While ice was dreaded it was safe and supreme in its isolation. But that which was once regarded as treacherous, useless waste is now a treasure, and is no longer sovereign of its own realm. Thick fleets of icebergs still guard the coasts where glaciers slide bodily into the water or spread wide over bays in forward-moving ice shelves, but now those defenses are pierced by ships with reinforced steel hulls, guided by air reconnaissance, radar, and other detecting instruments.

Icebergs are losing status. No longer do icebergs have freedom and supremacy of the seas with license to drift where they will, immune from attack. They, like glaciers, are slipping to the lesser rank of "usable resource."

The old daydreams about bringing icebergs from their home waters to warmer latitudes are being looked at in a practical light, for many coastal areas desperately need more fresh water than available sources can provide. For many years proposals to import icebergs were treated as a joke, but lately they have been given respectability.

In 1966 a long-range projection of United States' activities in Antarctica, published in the *Antarctic Journal of the United States*, contained this statement:

Despite engineering and transportation problems of enormous magnitude, there is a persistent interest in the possible utilization of icebergs as freshwater supply for rain-deficient areas. In the long term, the development of practicable schemes for utilizing Antarctic ice must be considered.

In the same year President Johnson submitted to Congress a broad program to meet future water needs and stated, "We must continue our search for bold, new ideas." In this program he urged study of the possibility of towing Arctic icebergs to southern California.

In 1969 Grigori Avsyuk and Vladimir Kotlyakov, Russian geographers, described how icebergs could be used as water supplies for coastal cities:

Even a relatively small iceberg (two kilometres [1.2 miles] long, half a kilometre [0.3 mile] wide and 150 metres [164 yards] thick) contains about 150 million tons of water—enough to supply each person in a city of eight million inhabitants with 1,000 litres [1,057 quarts] of water a day for a month. Towing the icebergs and tapping the water during the melting of large masses of ice would obviously present major technical problems, but theoretically the operation is feasible.

And at an eighteen-nation hydrology symposium Wilford Weeks of the U. S. Army Cold Regions Research and Engineering Laboratory seriously suggested that Antarctic icebergs as tall as a skyscraper and as broad as an airport could be pulled by giant tugs to the desert shore of northwestern Australia or the Atacama Desert of west-central South America, and there beached; and the melted water could be pumped inland.

There have been various iceberg-towing discussions of this kind over the years. The doubter's first thought is, the ice would melt before it reached its destination, but scientists who propose importing icebergs reason differently.

Just naturally and idly, icebergs often drift well into the subtropics, sometimes almost into the tropics. South Atlantic icebergs from Antarctica have been sighted within 1,700 miles of the equator. Some of the large North Atlantic ones last long enough to drift 200 miles south of Bermuda, which lies at 32° 20′ N, 2,230 miles north of the equator. However, most North Atlantic icebergs melt within a week or two after they enter the warm Gulf Stream. Currents normally carry them about ten miles a day, but sometimes as much as forty miles a day.

Most icebergs of the North Atlantic come from about twenty large tidewater glaciers on the west coast of Greenland. These outlet glaciers overflow from the ice cap and where they reach the sea bergs calve off in multitudes each summer. Smaller glaciers of Canada's Arctic islands supply Atlantic icebergs too but in far fewer numbers.

The icebergs have varied life histories. Many of the North Atlantic ones do not travel far. Thousands are stranded on beaches and melt there. Others are caught in the main currents and are borne directly into warm waters, where they melt. Some enter the Gulf of St. Lawrence. Some stay in their home inlets and straits a year or more—being frozen in place in winter—before being washed out to open sea.

If icebergs can have that long a survival record when steered erratically by winds and currents, they could retain much more of their bulk than they do on reaching middle latitudes if brought there quickly and directly.

If you look at an atlas map showing the location of deserts and cold ocean currents you will notice that most of the world's coastal deserts have cold currents offshore which come from a polar direction. Cold currents stabilize air masses passing over them and so are partly responsible for the lack of rain along those coasts. If icebergs were to be brought from polar or high middle-latitude regions to those desert shores they would insofar as possible be towed or pushed in the cold currents whose movement would carry them along and whose coldness would slow their melting. Eighty-five per cent or even up to 90 per cent of the berg would, of course, be under water. Much melting would take place in transit, but if the berg were large enough there would still be a considerable quantity of ice left by the time it reached port. Its exposed top could be given an insulated cover of some sort along the route, and fog, which is commonly associated with the cold waters, would shade the ice much of the time.

Icebergs from glaciers of Alaska (though not as large as polar ones) could be brought through the cold California Current, which flows south along the west coast of North America from just south of Alaska to Mexico, bringing solid water to arid coasts of southern California and northwestern Mexico. The cold Peru, or Humboldt, Current flows from the Antarctic Ocean to the equator along the western shore of South America, helping create the coastal deserts of northern Chile and Peru, including the Atacama Desert, the driest place in the world. Colossal icebergs from Antarctica could be towed north in this current.* Following the same circulation pattern, the cold Benguela Current in the Atlantic flows north along the desert coast of southwestern Africa, which could also use icebergs from the not-too-distant Antarctic Ocean. (South Africa has seriously considered using Antarctic icebergs to fill both industrial and domestic water needs.) Cold currents also reach the dry coasts of western and southern Australia. The cold current off southern Argentina could bring icebergs to dry Patagonia. In the North again, the cold Labrador Current flows from the iceberg-forming regions of Greenland and the Arctic Archipelago on south along New England and the Middle

---

* It was recently reported that Chile plans to tow icebergs to Antofagasta, a port near the Tropic of Capricorn. It is figured that even if the icebergs lose half their volume in transit the water they provide will be cheaper than that produced by desalination.

Atlantic States, and would be a favorable carrier of icebergs to cities
of the Northeast which, even if they are not desert, still need more
water, particularly more clean water, of which icebergs are made.

Small, densely populated islands whose area is too limited to collect enough water for its people, would be another good market for
icebergs. For example, Hong Kong, which rations its water, provided
the bergs could make it through the tropics.

How the meltwater would be collected once an iceberg arrived at
its destination is something for ingenious minds to work out. Small
bergs could be shoved onto shore or into a holding basin such as a
diked-off bay, enclosed fresh-water cove, or lagoon; but large ones
could not come to shore because of their underwater bulk. However,
they could be broken into small chunks and conveyor belts could
bring these onshore. Bergs might also be surrounded by a floating dam
which would trap the meltwater. It, being fresh, would float on top
of the denser salty sea water and could be pumped off into tankers or
reservoirs or the local water system. Water could also be drawn
directly off the ice as it melted. It would collect in natural or dug
hollows on the iceberg. In the old days whalers used to board icebergs
and fill their casks from pools of meltwater. Tunnels or gullies could
be cut in the ice too to channel the runoff to collection points. A
bonus of extra fresh water would accrue from condensation of atmospheric moisture on the ice.

Icebergs come in all shapes and sizes. In most areas they have irregular, jagged forms and melt down, of course, until they are mere
humps in the water. Some in the North Atlantic tower as high as
several hundred feet above the water. Around Antarctica flat-topped
bergs are more typical, for there pieces of level ice shelves break off
in profusion. Because of the way they form these bergs can be considerably larger than those found anywhere else. More than one third
of Antarctica's coastline is fringed with ice shelves, extensions of land
glaciers that stretch out over the sea. When pieces are broken from
those shelves by rough seas they are tabular and quite stable, at
least at the start. Icebreakers nudge some small ones around like
barges. The Ross Ice Shelf, the largest shelf, may at times crack off
in flat pieces about 600 feet thick and up to 100 miles across. Sometimes there are icebergs in Antarctic waters as large in area as Delaware or Connecticut. They drift around the continent for years.

Importing of icebergs could conceivably be a profitable business.

Several years ago a member of the Scripps Institute of Oceanography in California figured that an iceberg could be towed to Los Angeles from South Pacific waters by using six ocean-going tugboats of 80,000 horsepower. The cost would be about a million dollars, but the meltwater—enough to supply Los Angeles for a year—would be worth a hundred times that much.

The National Science Foundation is backing a more ambitious scheme now, to bring twenty-mile-long convoys of flat, perpendicular-sided icebergs from Antarctica to California covered with quilted water-in-plastic insulating film. Eight or more bergs could be linked by cables. Each berg would be about two miles long, a mile wide, and 900 feet thick, and would contain more water than a city of 750,000 uses in a year. The lead berg would have propellers and electric motors which would draw power from an escort ship. This convoy would need about as much horsepower as an aircraft carrier. During the trip of ten months, it is said, only 10 per cent of the ice would melt.

An iceberg parked in a harbor would provide more benefits than water. It would be a prime tourist attraction. And the ice could be used in many ways.

One can foresee the airlifting of small chunks of iceberg by helicopter from a harbor to polar-bear and seal enclosures at the zoo, or to anyone willing to buy them. Resorts might have some dropped in their lakes. A farmer might buy some to fill a dry pond. Any number of people would undoubtedly pay to have a shimmering little iceberg set in their private swimming pool to liven a splash party; or during a heat wave, to have one put in the yard where the kiddies could climb on it and where it would soak the dry lawn. The ice chunks would drip a little while being transported through the air, but the drips would be just like rain.

It has even been suggested that stocks of meat could be frozen in icebergs, kept in cold regions, and towed when needed across the oceans to be delivered to consumers.

One reason icebergs can stand a long trip better than would be expected is that glacial ice is much colder and melts more slowly than the ordinary ice we know. Its coldness is one of several attributes that make it especially desirable.

Men stationed in Antarctica discovered years ago that glacier ice added zing to their beverages and they would make special flights to certain glaciers to get their favorite brand of ice. Sailors and others who have sampled iceberg or glacier ice also know this. Now people back home are learning. It is already a gourmet item and a specialty in some stores.

Because glacier ice is colder than local or manufactured ice, if not allowed to warm in storage, it makes beverages colder, keeps them cold longer, and dilutes them less than ordinary ice. But that is not all. It is fascinating to use because of its oldness. It sings and crackles in one's glass or the punch bowl as it is releasing long-compressed air bubbles that have been locked in its cavities since the snowflakes and firn granules that formed it were packed together, maybe thousands of years ago. It is pure, free of DDT and other chemicals and pollutants. It is even tasteless because it lacks the minerals that running water picks up from the land.

The operator of rustic, isolated Taku Lodge, perched beside the Taku River of Alaska where glaciers spill off the Juneau ice cap, showed me how he motorboats up to the icebergs to chip off ice for his lodge guests because they greatly enjoy the novelty of having it in their cold drinks. Resort operators elsewhere probably do the same.

At Mendenhall Glacier near Juneau, Alaska, 200-year-old ice is "harvested" from floating icebergs for freezing fish and for cooling "Mendenhall cocktails." It is four times harder than manufactured ice cubes. On one occasion a block of this ice was flown to New York to ice the drinks at an annual Explorers Club dinner.

Other groups have also used glacier ice for their festivities.

Now ice chipped from North Atlantic icebergs is packaged in plastic bags and sold as party ice.

Icebergs no longer enjoy their former regal isolation safe from human molestation but are regarded as a deterrent to boating and shipping, mainly in ocean waters, that needs to be dealt with.

The world's most heavily used transoceanic shipping lane is that across the North Atlantic between eastern North America and western Europe. Each spring about 400 icebergs drift south past Newfoundland into that lane. They are a deadly menace, especially in fog or at night, and many tragedies have resulted as ships struck them. Radar is only partly effective in locating icebergs because they give off distorted images and because many of them are low. Their submerged part often extends out far beyond the visible part that bobs above the water line. In recent years there have been no collisions in this area, thanks largely to improved detecting devices and the vigilance of the U. S. Coast Guard, which keeps tabs on every iceberg in or approaching that shipping lane and advises ships of their positions.

Icebergs that hamper shipping would be destroyed if it were possible to do so, but they have a curious ability to survive attack. They have been shot at with cannon. They have been dynamited, bombed, torpedoed, struck with mines and heat bombs—all with no success. Nuclear bombs would do the trick but cannot be used, for they would contaminate the waters for marine life, and the Grand Banks off Newfoundland happens to be one of the world's richest fishing areas. Heat could destroy them, but it is estimated that two million gallons of burning gasoline would be needed to melt a medium-size iceberg.

A French scientist, Pierre-André Molène, believes icebergs can be destroyed by detonating an explosive inside them, blowing them apart from within. A torpedo with a heated nose cone would be lowered by helicopter. It would burrow deep into the iceberg, far below water level to a depth where it would have the greatest destructive effect, and then a delayed-action mechanism would detonate it.

Some consideration has been given to darkening icebergs with lampblack to make them melt faster, and this may yet be done. If some practical iceberg-destroying method should be found it will be good-by to icebergs in the shipping lanes of the North Atlantic. For now they are just kept under surveillance. But icebergs are beginning to suffer indignities much like those perpetrated upon wild birds and animals. They are being tagged for identification. The Coast Guard keeps track of individual icebergs by marking them with high-intensity dyes of various colors. They are shot at with arrows that leave a bleeding splash of color staining them and they are hit with color bombs.

When icebergs were unassailable and free-sailing, back around their home bases they successfully guarded the glacier strongholds from which they were launched, making them impregnable. Now icebreakers have a limited but growing ability to move through. We even talk of pulling icebergs thousands of miles to places where we want them, and we chip away at them for profit and pleasure.

Icebergs were almost dragged into World War II. There was a rumor that Germany planned to prevent invasion by jamming the English Channel with icebergs, but that story proved to be unfounded. Another even more unbelievable story was true, but the plan was kept secret until the war ended. The Allies were considering using icebergs for aircraft carriers.

In the fall of 1942 the war zone covered a discouragingly wide area, the Allies lacked air power in battle areas far from secure air bases, and submarines were an ever-present threat. The Allies' carrier-based planes were slower and less well armed than the enemy's land-based planes, so if the Allies were to invade distant areas they would be at a disadvantage until they managed to build airfields there, unless they had aircraft carriers capable of handling regular land-based aircraft. But such carriers would have had to be gigantic. Their cost would be tremendous. And at that time metal and other needed materials were in critically short supply. In October one Geoffrey Pyke proposed in a memorandum to the Chief of Combined Operations that large icebergs be used as aircraft carriers. Winston Churchill directed forthwith that research on that project be hurried and given highest priority. The icebergs were to be leveled for runways on top and hollowed inside to shelter the planes. The cost would be only a small fraction of the cost of constructing regular carriers, and no strategic metals or other material would be needed. A bergship would be protected with insulation against melting until it had fulfilled its mission and then a new one would replace it. Besides being inexpensive and as roomy as needed, the bergship had one great advantage over any man-made carrier. It could not be sunk! Icebergs, as we have noted, are extremely difficult to damage with explosives and, of course, unlike metals, ice will always float.

Bergships were never used in the war. Perhaps one reason was that most North Atlantic icebergs—those most easily obtained—are far from level on top. Methods of manufacturing icebergs in the shape of carriers were explored but were too complicated. Ice floes, or ice islands, were considered. These are the broken-up sections of the ice layer that freezes across the polar ocean surface. But they proved to be too thin for the planes to operate from. Had the Allies had the familiarity with Antarctica they have today, the ability to maneuver in its waters, and the bases there that they now have, perhaps the stable, flat-topped icebergs from those waters might have been given trial use. Their surfaces are somewhat undulating, but they could be smoothed with far less trouble than the irregular-topped bergs characteristic of the North.

The time is at hand when icebergs will no longer be secure in their own realm, able to frighten trespassers away. They will be one more of Earth's inspiring and beautiful decorations that humans dispose of when not wanted or use as they will. Some distant day perhaps it will be a rare treat to see anywhere icebergs as numerous and powerful as now. If more glaciers recede from the coasts and so melt before reaching tidewater, less ice will be cast into the sea. Developers of cold regions and environment modifiers may clear them from waters they want opened by blasting or blackening them or by artificially warming the waters or the whole area. Future generations may think back to this near-glacial period and wonder what it was like when icebergs clogged the gelid seas and sailed forth into strange waters like splendid galleons, never to return to port. They will have to merely imagine what they were like, as Samuel T. Coleridge did when in 1798 he told in "The Rime of the Ancient Mariner" of a sailing ship venturing into the little-known seas near Antarctica:

> And now there came both mist and snow,
> And it grew wondrous cold:
> And ice, mast-high, came floating by,
> As green as emerald.

And they can relive the sight and sound of icebergs in their original wildness and profusion in the writings of those who beheld them firsthand.

Admiral Richard E. Byrd wrote this description of his ship's precarious penetration of one of the most ice-packed sections of Antarctica's perimeter, the Devil's Graveyard. The time was late December, summer there. The year, 1933.

For days a sleet-oozing gripping oppressive fog, so thick that the bow at times was lost from the view of the bridge lay over the sea. Through that smoky pall like a phantom fleet, prowled icebergs past numbering with the sea throbbing in their basement grottoes. Like a cornered thing the ship stood amongst them, stopped and drifting, or manoeuvring with swift alarms and excursions to evade towering cliffs emerging with formidable clarity out of the gloom which bore down upon her. . . . A more God-forsaken place could not be imagined . . . our position is really critical. Visibility at times dropped to zero and the bergs rise out of the fog with alarming suddenness. You stare into the shifting vapour seeing nothing only dim uncertain shadows. Then cliffs upwards of 150 feet high cleave

through scarcely four ship-lengths away. Then the cry floats down from
the crow's nest echoed almost simultaneously from the lookout in the
eyes of the ship "Berg on the port bow." Down in the engine room the
engineers on watch spring to their post, the screw turns and the ship sheers
away from the menace. This darting to and fro has been going on all day.

On Christmas Day the fog began to drain away, and he continues:

Next day the sun broke through and all aboard beheld what is probably
the most unique sight on earth. . . . The sea was almost lost for the bergs.
There was no end of them. They rose one upon the other like the sky-
scrapers in a metropolis. . . . In the evening the sun strove to clear again
. . . and all the delicate latent lights—the fine blues, the lavenders, the
greens, were awakened in these floating creations. You saw whole cliffs
glow with the rare lovely beauty you associate with the light falling through
old stained glass in ancient cathedrals. . . . The Devil's Graveyard was very
different from the death prowl it had become in the fog.

John Muir, when he first canoed into Alaska's sequestered Glacier
Bay, viewed icebergs that were also solitary and bleak but almost
friendly in comparison to Antarctica's. This was his impression:

When sunshine is sifting through the midst of the multitude of icebergs
that fill the fiord and through the jets of radiant spray ever rising from
the tremendous dashing and splashing of the falling and upspringing bergs,
the effect is indescribably glorious. Glorious, too, are the shows they make
in the night when the moon and stars are shining. The berg-thunder seems
far louder than by day, and the projecting buttresses seem higher as they
stand forward in the pale light, relieved by gloomy hollows, while the new-
born bergs are dimly seen, crowned with faint lunar rainbows in the up-
dashing spray. But it is in the darkest nights when storms are blowing and
the waves are phosphorescent that the most impressive displays are
made. Then the long range of ice-bluffs is plainly seen stretching through
the gloom in weird, unearthly splendor, luminous wave foam dashing
against every bluff and drifting berg; and ever and anon amid all the wild
auroral splendor some huge new-born berg dashes the living water into yet
brighter foam, and the streaming torrents pouring from its sides are worn
as robes of light, while they roar in awful accord with the winds and waves,
deep calling unto deep, glacier to glacier, from fiord to fiord over all the
wonderful bay.

Already it is too late to see Glacier Bay as John Muir saw it then, for there are fewer glaciers there now, and fewer and smaller icebergs. What has happened there to weaken the glaciers' hold upon the land could happen worldwide if climate continues to warm, if humans develop the ability to modify climate as they are intent upon doing, and if they continue refashioning the earth's surface and using its resources for greater profit and the easier life.

Sometime if you are at a party enjoying a beverage chilled polar-cold with ice from an iceberg or glacier, listen closely amid the din of gay conversation to the soft "singing" of bubbles being freed after compression and imprisonment of hundreds or thousands of years, and hear the warming ice crack under stress for the last time. Pause among the frivolity and think how those tiny pockets of air were caught among snowflakes that fell on unknown lands long ago, maybe while far away Cleopatra entertained Caesar and Antony, or Solomon reigned, or the Trojan War was fought. Or still earlier when rugged forebears of yours hunted reindeer on a tundra in central Europe or woolly mammoths in North America, or fished in now-vanished Saharan lakes.

How many others besides you—as they savor a last delicious piece of ice, press it between their tongue and palate, and feel it melt to nothing—will be thinking of all this and of the magnificent Ice Age dissolving into memory?

CHAPTER 22

TOURIST EROSION

O for a lodge in a garden of cucumbers!
   O for an iceberg or two at control!
O for a vale which at mid-day the dew cumbers!
   O for a pleasure trip up to the Pole!*

ROSSITER JOHNSON

A GEOGRAPHY textbook lists the principal agents of erosion as water, ice, and wind. As they move they pick up material and carry it along and so wear down, eat away, features on the earth's surface. One could add a fourth agent—human beings. Now numbering in the billions and equipped with all manner of machines and tools for scratching, scraping, and moving things, we are one of the world's busiest erosive forces. Besides actually transporting materials we loosen them and so facilitate the job of the natural agents. The capabilities of farmers, miners, land graders, road builders, and constructionists are obvious but there is another subgroup of human eroders that is rapidly increasing its range and effectiveness. Tourists.

Tourists are usually thought of as harmless, gentle folk who just sight-see but do not touch. This, of course, is a fallacy. They . . . we . . . are some of the greatest touchers, pokers, picker-uppers, and carriers of all time. Adoring tourists can transform and mutilate a place without meaning to, not only by unavoidable wear and tear and souvenir collecting, but because facilities must be built to get them to their destination and administer to their needs.

In coming years there will be more tourists than ever before, due to

---

* From "Ninety-nine in the Shade" by Rossiter Johnson from the *Book of Humorous Verse*, edited by Carolyn Wells. Copyright 1936 by Doubleday & Company, Inc.

the population explosion plus the fact that more people have more time and more money to travel. Workers are being given shorter workweeks and longer vacations, and are retiring earlier in life. Travel is becoming cheaper and easier, and one can go long distances and back in a few days.

Already travelers are wearying of the old sights, the musty museums, the overcrowded resorts, and are seeking new places to visit. Many discriminating tourists do not care to go where there is political unrest or disease, no matter how colorful the travel folders are, or to places where the pace is as fast, the cacophony as loud, and the entertainment as bawdy as the place they paid to escape from. They want to find uncluttered, broad horizons, peace of mind, and to return home inspired and refreshed.

And so the glaciers beckon, and the sharp-crested mountains where they used to lie, and the cool, jewel lakes whose basins they formed, and the mossy, woodsy land they left not long ago. The restful, the quiet, the clean, the green, even the bare rock, and especially the white ice and snow. Vast, untouched canvases on which one can paint beautiful mental pictures with one's inner thoughts.

Glaciers are, indeed, one of the world's most dazzling, inspirational, and ancient wonders, with architecture, history, and splendor beyond compare. Those near inhabited areas have been popular recreation and health-resort sites since Western tourism began around the start of the last century. Recently distant ones have become accessible too. What greater thrill for a traveler than to go where few others have been and to feel like an explorer instead of a tolerated guest.

If I am promoting travel to glaciers I regret doing so, in a way, for I do not want them to be desecrated like every other lovely place that people swarm over. Still, I wish everyone could feel the excitement of being in their presence. The pity is that when the glacial sanctuaries are opened to large influxes of visitors the glaciers and their natural picture-frame settings of polished rock or velvet pastures or tundra meadows or evergreen trees will be spoiled by what might be called "tourist erosion."

Tourist erosion takes many forms. One kind occurs in famous old buildings and ruins where parades of sight-seers pick off pieces to take home. So the edifices crumble a little faster. Natural features are no safer. Interesting rock formations everywhere have been chipped at by admiring, sampling "nature lovers." Arizona's Petrified Forest, for example, would have "walked away" completely if restrictions had not been imposed against taking the petrified wood. One would not expect anyone to take pieces of glaciers with him, and yet this is done, sometimes just for fun, sometimes to pack a catch of fish or to cool a camper icebox or for various other uses. Occasionally the free-and-easy consumption of glacier ice reaches commercial proportions. It is, after all, colder and longer-lasting than other ice.

One place actually encourages tourists to take its glacier ice along. As you might guess, that is Greenland. The Consul General of Denmark, to which Greenland belongs, informed me that any traveler in Greenland is allowed to take away "as much as he can carry of this rare commodity." Greenland's most abundant resource, which used to be considered nothing but a curse, is in fact becoming a tourist attraction and exportable item. Denmark has distributed to foreign markets samples of Greenland's inland ice in cube form to stimulate business. "Ice-cap cubes" they are called, guaranteed to be at least 2,000 years old and absolutely pure. Commercial airlines considered using Greenland ice a few years ago but getting it was too great a problem. Greenland's ice cap is not hurt by ice exports, of course. One of its large outlet glaciers alone dumps more than 20 million tons of ice into the sea on a summer day. But small, shrinking glaciers elsewhere in the world could suffer serious attrition if ice were taken from them in quantity. And their melting would be hastened if their surfaces were eroded or darkened, or if their surroundings were made warmer. Such artificial changes are taking place.

Sometimes glacier ice is requested for a certain festive occasion. It may be sent considerable distances. For instance, the Overseas Press Club of America periodically holds dinners in New York at which food typical of a certain country is served, and when its New Zealand dinner was held it imported from New Zealand not only the food and beverages for the occasion but also the ice. It was cut in three-foot-square chunks from Tasman Glacier in the Southern Alps at about 7,000 feet, flown to Auckland, put in the freezer of an ocean liner going to San Francisco, and then flown to New York. It was to be used in drinks, but because tests showed it had acquired human bacteria in handling, the New York Board of Health said "no" and it was used only to set a punchbowl and dessert dishes in.

A glacier may be eroded in just a figurative sense by having its natural patina marred, or it may be eroded quite literally, having its bulk reduced. Its setting too—the ring for the gem—may suffer deterioration. A once-wild glacier deprived of its proper setting, with its natural appearance altered, its muscle weakened, is as deplorable as a caged animal in a zoo.

The most vulnerable glaciers are those in climatically marginal areas which have warm enough summers to cause appreciable melting and which, because the climate is hospitable even in winter, have been visited by growing numbers of people who like that fresh environment and who want beautiful playgrounds and virile recreation. Glaciers are not protected or respected. Many a sick one is exploited for profit, entertainment, or sport. In some cases the glacier's melting may be inadvertently speeded, and maybe its demise. It may hold its own as long as it has a clean, snowy surface and good reflecting ability, but may be in trouble when beset by tourists. Its surroundings are apt to warm as the shady forests and tundra mat are cleared to make room for roads, airports, asphalt parking lots, hotels and cottages, sanitary facilities, camp sites, lookout points, trails, ski runs, heated swimming pools, golf courses, and other recreational areas. Its mirror surface may be darkened by dust sent flying as ground and rock are dug and blasted up; and by smoke from buildings and campfires, and exhaust from cars, tour buses, trucks, airplanes, and snowmobiles; and by the muddy tracks of vehicles that drive right on the ice. As the ice is soiled it melts faster. Also, a haze of impurities in the air, caused by the new settlement, may hold in heat and raise the air temperature.

Large, enclosed snowmobiles have for years been carrying whole groups of tourists onto a glacier at one time. Fleets of these drive here and there over the ice. Passengers get out on the glacier and walk about to inspect the scary cracks and gurgling holes, and with some disappointment they see that the glacier's natural look is worn off. The vehicles' biting treads have cut the glacier's surface into slush and ruts. Periodically scraping machines plane it smooth enough to drive on again, causing further erosion.

Each year more thousands of individuals are operating their own smaller type of snowmobile for sport or work. This vehicle has become so popular that there are not enough public trails to accommodate the traffic. Snowmobilers (many of whom do not respect fences or restrictions) are bound to want to use the wide-open glaciers for trails and racetracks.

Ski planes and helicopters land on and take off from glaciers, bringing tourists up and down. Hikers and pack trains of animals wear tracks over glaciers, and paths are carved through rough places. Youngsters play and sled on the safe slopes. Mountain climbers find it easier and safer to hike up and partly slide down a gently sloping glacier where possible—guarding against crevasses and holes—than to scale and descend steep, jagged rock cliffs. Many climbers do not bother starting at a mountain's base but are flown halfway up, often onto a smooth glacier, and there they make camp and start their climb. Needless to say, with this saving of time and effort the sport of climbing mountains and glaciers is rapidly acquiring new participants. Streams of skiers in resort areas cut tracks in the snowy protective cover of glaciers day after day, month after month. Even night skiing is common, and a necessity in places because of the crowds. The wearing effect may be called insignificant but it could grow as sports activities and tourism focus increasingly on glaciers.

Tourist boats are taken as close as safely possible to the fronts of glaciers entering lakes or the sea, and passengers wait with bated breath and cocked cameras to see icebergs calve off and splash into the water. When the glacier does not oblige or delivers its ice too slowly to suit the customers, the boat operator makes the glacier crack off faster by sending vibrations in its direction. He lets off blasts from the boat's whistle or fires a gun. Sometimes he fires directly into the glacier. As chunks of the glacier front topple down prematurely, creating spray and frightening waves, passengers cheer and hurriedly take their pictures.

Not until the middle of the nineteenth century did it become generally known that glaciers move. Before that they were regarded as just ice that accumulated in mountains. (Continental ice sheets were not yet envisioned. Newly discovered Antarctica was just being peeked at around the edges. Greenland was not crossed until 1888.) Now people became fascinated with glaciers and wanted to see these creeping marvels for themselves. When Mark Twain toured Europe in 1878 and wrote of his experiences in A Tramp Abroad he gave considerable attention to glacial phenomena, then a subject of much interest. Already glaciers in the Alps were being exploited to attract customers, as recorded in his telling of a visit to the Glacier des Bossons on Mont Blanc.

One of the shows of the place was a tunnel-like cavern, which had been hewn in the glacier. The proprietor of this tunnel took candles, and conducted us into it. It was three or four feet wide and about six feet high. Its walls of pure and solid ice emitted a soft and rich blue light that produced a lovely effect, and suggested enchanted caves, and that sort of thing. When we had proceeded some yards and were entering darkness, we turned about and had a dainty sun-lit picture of distant woods and heights framed in the strong arch of the tunnel and seen through the tender blue radiance of the tunnel's atmosphere.

The cavern was nearly a hundred yards long, and when we reached its inner limit the proprietor stepped into a branch tunnel with his candles, and left us buried in the bowels of the glacier, and in pitch darkness. We judged his purpose was murder and robbery; so we got out our matches and prepared to sell our lives as dearly as possible by setting the glacier on fire if the worst came to the worst—but we soon perceived that this man had changed his mind: he began to sing, in a deep melodious voice, and woke some curious and pleasing echoes.

Similarly today, tourists are brought into caves and tunnels of glaciers. Their body warmth, the mixing of the air which they cause, and the heat from the lighting raise the temperature there inside the glacier.

Mark Twain encountered more tourist erosion on the famous Mer de Glace, named for its billowy surface, which was like a sea of ice whose waves were frozen. The eroder was a man who for a price chopped steps in the glacier to help tourists walk up the slippery bulges. This glacier, like many others, was a social center. Hundreds of people came in a day. Mark Twain said of the Mer de Glace, "There were tourists of both sexes scattered far and wide over it everywhere, and it had the festive look of a skating rink."

For many years mountain railways, aerial cable cars, and chair lifts have been taking sight-seers and skiers up Europe's Alps to the glaciers, and now planes too are bringing them up, as they are in many tourist areas throughout the world. At Switzerland's highest railway stop—Jungfraujoch, 11,203 feet—is an "ice palace" where entire rooms are carved from a glacier that has been stationary (that is, barely holding its own) for a long time, and guests are given rides over a glacier in a sleigh pulled by huskies. There are other such "ice palaces" in the Alps.

Europe's glaciers abound with ski runs and practice areas and various forms of entertainment. Norway promotes skiing competitions on a number of glaciers every summer.

The tourist trade is discovering new glaciers that used to be undeveloped or beyond their reach—in Iceland, Spitsbergen, southern Chile and north through the Andes, the Himalayas and other mountains of Middle Asia, on Baffin Island which faces Greenland, and elsewhere. Glaciers are a strong attraction wherever they exist and can be reached, so promoters are making it increasingly easy for even unathletic travelers to reach them.

In New Zealand ski planes have been landing tourists and skiers on Tasman Glacier on South Island for many years. Tasman Glacier is on Mount Cook, the country's highest peak—12,349 feet. From the landing point one can ski nearly eighteen miles down the glacier's valley. The airplane ski lift is a regular service, flying up thousands of passengers per season. Not all ski back down. Anyone can fly up to the glacier, scamper around awhile, and fly down again. On busy days more than 300 tourists have been lifted to the glacier. With small, five-passenger planes, that amounts to more than sixty landings and take-offs a day on the glacier, and business is sure to increase.

British Columbia is one of many regions where helicopters have become popular lifts, including fifteen-passenger jet models. They carry glacier skiers to the top of beautiful, high, trackless runs that they could not have reached otherwise. Entrepreneurs from Austria, Switzerland, Germany, and New Zealand are developing some of this spectacular glacier country.

One can foresee the day when commercial tourism will be as highly developed in the glacier regions of western Canada and Alaska as in the Alps. Since the Alps are now saturated with tourist facilities, these new, untouched areas, especially those with larger, more impressive glaciers, seem likely to supersede old European tourist centers as favorite gathering places of many of the world's skiers and mountain lovers in years ahead.

Alaska has the largest, handsomest array of active glaciers outside the polar areas. It also has the world's flyingest people—the most planes per capita anywhere. This combination brings people and glaciers together readily. But there one does not even need a plane to come close to glaciers. One can hike or drive right up to many of them. And spectacular water-level glaciers are approachable by small boats or large cruise ships. From the capital, Juneau, you can reach the lake edge of magnificent Mendenhall Glacier by bus or taxi, then walk a trail to the glacier if you wish. Or stay snug in the glass-enclosed Visitors' Center, and be impressed at how far the Mendenhall has receded, for the site on which the Center stands was under ice until 1941. Since then the ice has drawn back a half mile. The lake in front of the glacier is used for ice skating in winter.

Worthington Glacier near Valdez, Alaska, is a favorite source of free ice. Tourists drive up to it and fill their camper ice boxes.

Glacier Bay used to be so isolated that only the more daring Alaska visitors braved the icebergs and mountainous desolation to see it. When John Muir went there in 1880 it was total wilderness and solitude, but only a few years later tourists were looking it over. In 1925 the bay was made a national monument, and up to about 300 persons a year visited the still-remote fiord. In 1960, 900 people visited Glacier Bay. In 1965 the National Park Service built a rustic lodge there, a most incongruous haven of modern comfort set in the wildness of mossy forest, bouldery slopes, and whale-inhabited waters. In 1970 attendance had risen to 29,700! Tourist business is rapidly on the increase there as planes fly in regularly—now only a half hour from Juneau—and boats ply up-bay to the last of the dwindling glaciers that once filled the whole inlet.

Bradford Washburn, director of Boston's Museum of Science, who has done much exploratory work and brilliant photography in Alaska, has deplored the fact that tourists cannot have easier access to Alaskan glaciers. The Associated Press reported a comment of his: "Vast fields of tidewater glaciers, tremendous mountains, and waterways of Alaska are unmatched in the world. Yet casual visitors are able to go just so far." He feels Alaska should have glacial tunnels, ice palaces, and tramways up mountainsides to satisfy tourists. He advises protecting some glaciers "in their total natural glory" but developing others as intensively as resort areas of the Alps. Alaskans may respond to this kind of urging. Many of them want their wilderness areas developed rather than preserved. The state, which holds the last of United States' totally wild and robust glaciers, is not yet as conservation-minded as those that know from experience how natural beauty can be swiftly and irrevocably lost.

Glaciers closer to population centers are in more immediate danger of tourist erosion. In the "lower 48" the healthiest concentrations of glaciers are in the Cascade and Olympic mountains of western Washington. A number of these glaciers are already accessible and many more would be with some means of air lift. There in the Westerlies near the sea, mountains receive exceptionally heavy snows and summers are cloudy and cool. The Olympic Mountains hold over sixty glaciers. The giant 14,410-foot volcano, Mount Rainier, in the Cascade Range, alone supports over forty, including the five-and-one-half-mile-long Emmons Glacier, largest in the United States outside Alaska. The Cascades of northern Washington sustain nearly half of all the glaciers of the West.

Since the North Cascades National Park was created in 1968 the National Park Service recommended putting several tramways in it and eighteen non-wilderness enclaves ranging in size from nine to thirty acres. The Wilderness Society stated its objections:

The proposed tramways would carry mechanical invasion of the wilderness to the mountaintops. The proposed enclaves would pockmark the finest natural wilderness of the park. They would contain all sorts of man-made structures including sewage and electric systems, helicopter landing sites, and chalets or hostels providing food and lodging. A more direct and effective way of destroying the entire concept of wilderness in the park is difficult to imagine.

The press is on. This is the constant dilemma: how to bring people into an area to enjoy it, and how to preserve the beauty they come to see which they, by their very presence, help destroy.

A study was done in Rocky Mountain National Park, Colorado, to learn what effect tourism had on alpine tundra vegetation along Trail Ridge. Indians had used this high trail across the glaciated crest for 7,000 years without hurting the vegetation, which consists of mosses and flaky lichens, and cushiony, leafy and flowering plants. But destruction began when the trail was opened to tourists. Paths were provided, but not all visitors stay on paths. Tourist trampling ruined the vegetation cover in some places in just one season. Where use has been heavy all plants are gone except a few protected by boulders, and five inches of topsoil has been worn off. Botanists judge that in some locations the tundra plant community will need over 500 years to restore itself.

Montana has an interesting glacier that tourists found reason to whittle at—Grasshopper Glacier in Custer National Forest. At one time millions of grasshoppers were embedded in it. It is believed that about 200 years ago they were migrating and became chilled or were caught in a snowstorm and dropped on the ice. Until recent years the grasshoppers were preserved, frozen in the ice. Visitors dug into the ice for specimens that had not decomposed through exposure. Birds and animals ate the insects. Lately thawing has exposed so many of the grasshoppers that they are becoming hard to find. Natural wastage has reduced this glacier's length from four miles to about one mile since it first attracted attention. There are other "grasshopper" glaciers besides this one, which is easily reached and therefore well known.

Under the heading "Preserving Natural Features" a National Park Service bulletin for guests of Glacier National Park reads: "It is against the law to disturb, injure, remove, or destroy vegetation, wildlife, rocks, or fossils within the park." It says nothing about glaciers, the feature for which this park in the Montana Rockies was named; and they are in greater peril of disappearing than any of the mentioned features. One can hardly see a glacier in the park any more, and if he does it looks like a little patch of snow on the mountain side. There are only some fifty glaciers remaining and most are only glacierets. All are on the north or east sides of high peaks and ridges between 6,000 and 9,000 feet. This is below the snowline now, and none could exist in this too-warm climate except that wind-drifted snow is blown into their basins. Only two have an area that approaches a half square mile—Grinnell and Sperry glaciers. Not more than seven others have areas over a quarter square mile.

Grinnell and Sperry glaciers, like the smaller glaciers, have shrunk rapidly during this century. Sperry was the larger in 1901 with 810 acres, but a 1960 measurement showed it to be down to only 287 acres. Grinnell shrank from 525 acres in 1901 to 315 acres in 1960. Between 1937 and 1946 its volume reduced one third. In 1946 a large part of its ice front fell into its marginal lake, completely filling it with icebergs. Grinnell is still said to be the largest glacier in the park, but actually it has divided into two small sections. Its central part melted away. The upper section now has its own name, Salamander Glacier.

Visitor attendance at Glacier National Park has increased steadily. In 1940 attendance at the park was about 177,000; in 1950, 482,000; in 1960, 725,000; and in 1970, 1,242,000. The main center of activities is Many Glacier Hotel (where glaciers are now conspicuous by their absence). While there I listened to a guide give an evening lecture, inviting park guests to join his hiki ᵀ tour to Grinnell Glacier the next day. He told how the trail to the glacier was opened at the start of the tourist season by blasting away ice and snow blocking the path. He warned that no one was allowed to pick wild flowers on the way up or deface rocks; but to encourage his listeners to take his tour he told how, when they reached the glacier, he would chip off some ice for everyone and made our mouths water telling how delicious it would taste. The disparity in values was not sensed.

One may not chip off a piece of rock in those, the highest Rocky Mountains in the United States, nor bother wildlife, nor pick a flower for fear of being fined, though flowers may be blooming everywhere we look. And that is as it should be, so that everyone who comes may see things in their natural state and so that rare specimens will not be wantonly destroyed. But why is it that glaciers—less permanent than rocks and just as defaceable as they, and unable to reseed, reproduce, or run for cover—are treated as expendable? If they are hacked at, if their habitat is disturbed, that is permitted. One chunk off a glacier may be proportionately as infinitesimal—or as great—a loss as a picked flower, but we do protect flowers.

It seems strange that conservation principles are applied to everything around glaciers but never to glaciers themselves. Glaciers are treated not as resources that could be lost, but as curiosities or shivery things whose preservation is of no concern.

Glaciers are composed of more than ice. They are transporting assorted earth materials, including rocks. Smooth, rounded rocks or those faceted or striated by glacial abrasion are kept as art objects or scientific specimens by many tourists and rock collectors. They are not scarce. Glaciers pile billions of them in all sizes around themselves. But there is something special about personally picking one up from the surface of a living glacier. I have a number of such rocks, and no harm was done in taking them. The glaciers were not hurt nor any law broken; and a geographer by nature and of necessity collects rocks. But this leads to two points. One: When scientists travel they are partly tourists at heart. Two, a question: Where is the distinction between scientist and tourist in regard to taking liberties with the landscape, and why does the scientist usually have more rights than the tourist? Because scientists must engage in research. Also, because presumably they can better appraise the value of, and the need to preserve, a given natural feature; and because it is expected they will be more careful. These assumptions are not always correct.

Therefore, one might include scientists among the perpetrators of "tourist erosion." We avidly pick up samples of rocks, fossils, soils, vegetation, artifacts, and other things to add to our collections, to do analyses of, to use in teaching, and occasionally for status symbols in our offices. Often specimens are enthusiastically collected and sent home from field camp by the cratefuls or truckloads, many never to be looked at again. Stored in basements and attics of university buildings, museums, and other scientific institutions are innumerable unpacked crates of collected materials that will never be examined or ever see the light of day again till the building itself comes down.

Most "erosion" of glaciers and glacial landforms that scientists do is essential in gathering data and instructing students. Effects of small-scale ice-melting experiments and drilling and digging in the ice are quickly erased and do not damage large and sturdy glaciers; but small, weak ones that manufacture barely enough ice to maintain themselves or that are rapidly wasting could suffer from induced ablation as more and more eager scientists and scientists-to-be are turned loose on them.

Earth scientists of the past are remembered in general as members of a small, dedicated corps, who would endure financial and physical hardships and would sacrifice long periods of time to do field research in hard-to-reach places. But today the number of scientists— and the term is used broadly—is legion, and with way-paid expeditions and rapid, comfortable, prearranged transportation they roam the world. In tourist style they hop on planes and with a minimum of effort land directly on glaciers and ice caps with their equipment. They—especially crews engaged in military and government operations—bring in heavy-treaded construction, grading, and hauling vehicles, machinery, and heating equipment, among other things, and go to work making their marks on the ice more strongly than ever before. They build their camps on the ice, and in Greenland

and Antarctica have hollowed out tunnels, rooms, and whole heated towns inside the ice sheets.

Scientific expeditions and research jaunts are not everywhere limited to highly trained or mature personnel. Helpers with various skills go along, including students. Glaciers have become popular teaching and research laboratories.

Glaciated terrain also is good educational field-trip country, and that which is near schools is flooded with students. This is desirable and required if all are to study these terrain features firsthand and acquire experience. But precautions have to be taken.

Students energetically chip away with their hammers or dig with their shovels because they each want their own specimens, or they want to photograph an unweathered surface, or to see for themselves a freshly broken rock face, soil profile, drift deposit, or bedding plane. And they are taught to do so. Digging and hammering in the field are essential to research and teaching. Most students are careful when forewarned and supervised, but occasionally this unnatural erosion is overdone to the extent that a valuable site is ruined for further research and for posterity. (On the other hand, it often happens that specimens collected and preserved by scientists—including amateurs—are ultimately all that remains of a feature that is ripped up by the march of civilization.)

The Ice Age National Scientific Reserve, a series of parks in Wisconsin, was established to preserve remnants of Ice Age terrain, distinctive features created by the ice sheets. One of the more dramatic settings is Devil's Lake gorge. Students by the busloads, tourists and campers, scientists from around the world, go there to enjoy and study the picturesque site. Those scientists who appreciate its importance have been concerned over the unnecessary "erosion" of irreplaceable evidence of past events. Glacial geologist Robert F. Black felt strongly enough about the matter to write in the *Transactions of the Wisconsin Academy of Sciences, Arts and Letters*:

It is urged that the many striking features illustrated in the Devil's Lake area be seen and appreciated, but not destroyed, by the thousands of visitors each year who come to Devil's Lake Park. Pressure of man's use continues to increase each year, now to the point where even the durable rocks need protection. In their zeal, geology students particularly have contributed disastrously to the natural attrition of certain exposures of the bedrock. Every geologist who has written extensively on the Devil's Lake area has emphasized the uniqueness of the glacial, periglacial and bedrock phenomena present. No other location in the midwest has such a rich variety of unique features in so small an area near major centers of population. As a tourist area and as the scientist's field laboratory, it is certainly unrivaled for hundreds of miles around. Hence, every effort must be made to preserve, not just the features in the Park, but the many glacial and bedrock features adjoining it as well for future use of all mankind. Once destroyed, they can not be replaced.

Also within the Ice Age Reserve is the Two Creeks Interglacial Forest Bed on Lake Michigan's shore, where on the open face of the lake bluff can be seen the fallen logs and stumps of spruce trees, radiocarbon dated at over 11,000 years of age, buried between layers of drift. It is evident that these trees grew up after one ice sheet withdrew and were covered by the advance of the next. Before this unusual exposure became part of the Reserve so much of the rare wood was torn from the bluff by visitors, many of whom were science instructors and their students—for they were the ones who knew the wood's importance—that further removal had to be prohibited to slow the bluff's erosion and leave some of the exposed ancient forest bed *in situ* for others to see.

The counterpart of "erosion" is "deposition." Any material that is lifted up must be set down, or what is eroded must be deposited. So we also have "tourist deposition." Usually this takes the form of littering, but it can occur in other ways. Recall that even with Man's first visit to the almost-sacred moon, his erosion (bringing back rocks) and deposition (leaving equipment behind) began immediately.

Lest one think high mountain summits could be forever safe from litterbugs let him cast his eyes up to Mount Whitney in California, which was United States' highest mountain until Alaska joined the Union, and see how far the droppings of our species have spread. The top of that 14,494-foot mountain is being called America's highest trash can. The erosion-deposition process normally carries material downslope, but here it is reversed, temporarily at least. Hikers up the thirteen-mile trail to the summit of Mount Whitney have left such a collection of litter up there that trains of pack animals must haul down tons of it each summer. Helicopters help carry it out, and the National Park Service is considering building a heliport on the summit just to facilitate trash removal.

Mount McKinley, North America's highest peak, which seems too regal and remote to be affected by litter, has gathered so much trash that a team of mountaineers from Oregon recently climbed it to clean up some of the debris left by other climbers.

The glacial record is being confused by some unnatural erosion and deposition. One example: A certain pilot who often flew to Antarctica bore a personal grudge against geologists. He used to pick up rocks in various parts of the world and then when he was in Antarctica he would drop these out-of-place rocks from his plane over glaciers on which he knew geologists were working in order to mislead and confound them.

Another example: Geology and geography professors have long been taking their classes to the Driftless Area of southwestern Wisconsin and parts of neighboring states, that "island" that escaped being run over by the ice sheets. Students know that on a field trip to that area the professor would be stressing the absence of glacial deposition. So before these trips it would often happen that practical jokers would haul boulders and smaller rocks from surrounding glaciated territory and place them conspicuously where they would be encountered during the field trip, to the embarrassment of the professor. No one knows how many of these glacial erratics have been scattered through the Driftless Area by pranksters. Some may sometime be cited by serious field researchers as "evidence" of glaciation there.

A strange kind of tourist erosion has been going on in some of Europe's caves that Upper Paleolithic artists decorated with magnificent animal pictures. When those caves were deserted and forgotten at the end of the Ice Age, those masterpieces were locked in darkness and preserved in constant temperature and humidity. No animal reproductions since have surpassed them, and their mystic, antique realism can never be duplicated. They are lifelike glimpses into the Ice Age which could not have been seen had these pictures not endured. It was natural for tourists to want to see them, and they came in great numbers.

The mayor of one town rented a local cave that had these Ice Age picture galleries, installed electric lights, erected a bar in a corner, and supplied dance music. Tickets were sold at the entrance. Customers tested the still-greasy colors of the paintings with their

PLATE 39. The digging away of glacier-made sand-and-gravel landforms which has been going on for years (notice the old-fashioned machine used to load railroad cars) has resulted in the removal of many of these features or has left them in an unsightly state such as this. Large unwanted boulders are piled aside.

PLATE 40. People viewing a mighty glacier are awed by the strength of its powerful grasp and forward thrust. But are they equally impressed by its weakness seen in the rivers of meltwater pouring from it? Only time will tell whether nature and mankind will allow glaciers like this to survive, or whether glaciers, like other vestiges of the Great Ice Age, will continue to disappear from Earth's scene.

fingers. Their clothes rubbed them as they passed by. Many drawings in many caves have been worn off the walls in this way.

To make the picture caves inviting to tourists many were provided with easy entrances, sidewalks, and steps, and artificial lighting illuminated the pictures.

But after some years the pictures began to fade and threatened to disappear, taking their ancient memories with them. Indrafts of outside air and the breathing of thousands of visitors altered the atmosphere of the caves.

In one of the most famous and beautiful of the caves, that of Lascaux, the pictures had remained in near-perfect condition for perhaps 20,000 years. Then it was opened to the public after World War II, and in less than twenty-five years the paintings were deteriorating. Each year 125,000 visitors entered the cave. A special air-conditioning system and bronze air-lock doors at the entrance were supposed to protect the paintings, but they did not. In 1963 ugly green blotches of fungus-like growth began creeping across the walls, eating the paintings. The cave was then closed to the public and hermetically sealed while ways were sought to halt the spread of the "green peril" that threatens to erode some of Ice Age people's most eloquent communications to the living world.

Altamira Cave, with its equally famous paintings estimated to be between 17,000 and 20,000 years old, was also opened to the public. A road was built to the cave; and stairs, cement paths, and a waiting room were provided. It was proudly said that the cave could be visited as easily as a museum, and each painting was illuminated to the best advantage with neon lights. Now, however, the pictures are fading from that lighting.

Let's look at Antarctica. Since military and scientific personnel, shippers, and other workers have been there, unnatural erosion and deposition have been altering that last virgin continent, although precautions are taken to see that that does not happen. The meager, ground-hugging vegetation is steadily retreating from occupied areas, and declines in marine and terrestrial wildlife populations are reported. The simple ecosystems so important scientifically and esthetically cannot endure much disturbance.

Antarctica's waters teem with animal life and the ocean floor is a fascinating community, but divers report that the underwater scene has changed markedly in the last few years, for the worse. It is com-

mon practice for garbage and trash to be hauled away from the bases
onto sea ice, where it stays until the ice breaks up, and then it falls
into the sea. On the floor of McMurdo Sound grew giant sponges,
some of which are believed to be a hundred years old. These are being buried and killed by debris dumped into the water—barrels, cans,
rope, cardboard, buckets, tractors, clothes, pieces of airplanes. Oceanographers say that the bottom of the Sound will soon become a veri-
table rug of litter if this method of waste disposal continues. But the
large amounts of food wastes and accumulated sewage added to the
water are even more destructive.

Palmer Station was for a time accumulating so much litter that it
was called "a prime example of land pollution" and a clean-up was
ordered. Hallett Station was even worse, for it "eroded" a penguin
colony as well as produced undesirable "deposition." There nearly
8,000 penguins were fenced from their nesting sites, and buildings
and antennas were erected in their rookery. Construction debris and
discarded equipment were piled about. Refuse dumped on sea ice
eventually washed ashore, and human wastes and garbage added to
the problem. Steps were taken to improve conditions there.

At first military and scientific personnel had the big ice sheets,
Antarctica and Greenland, almost to themselves, but now tourists
are making their presence felt there too.

Why would tourists want to go where it is bitter cold? Is that enjoyment? It is, for people who travel to learn, to see the world and
feel things as they are, for pictures alone will not satisfy them, nor
all the descriptions in travelogues or geography books. They want
to experience a place as it is, frigid and blizzardy if that is its normal
state. These ice continents are not disagreeable all the time, any
more than winter is. There are times of stillness and sheer beauty.
But the icy coldness and unchecked wind are the essence, and you
would not wish to see Greenland or Antarctica without feeling them
any more than you would want to visit London and not see the fog.

Tourist accommodations in Greenland are still modest but already
there is a hotel of sorts at Narsarssuak. One stops at busy coastal villages and sees traces of old Norse settlements. Fishing is great and
camping in tents a real adventure. And, of course, there is the profusion of ice.

Tourist cruises began going to Antarctica in 1958 from Argentina.
Now group tours originate in many countries in both the Northern

and the Southern Hemisphere. Tours fill up fast. Applications come from all over the world, even before dates and prices are set. Tour members from the United States have been flown first to southern South America, New Zealand, or Australia, and taken from there to Antarctica in a ship which serves as their hotel, for there are no tourist accommodations on that continent. An eighty-six-year-old woman went on one tour. New Zealand plans to fly tourists directly in to "the ice," as Antarctica is referred to, in order to handle more customers because the tourist season is short, and besides the ocean trip through the "roaring forties" is rough; but until lodges or hotels are built—as they ultimately will be—passenger liners will be moored on the coast during the tourist season and used as floating "boatels." They will probably be docked near the United States base at Mc-Murdo, the largest permanent base in Antarctica, and the tourists could headquarter there a week or so and fly back. Christchurch in New Zealand has been a jump-off point for many scientific expeditions to Antarctica. Planes can fly the 2,400 miles from there to Mc-Murdo in a few hours.

Tourists are taken only to places along the edge and northernmost extension of Antarctica, and they go during the warm and daylight season, our winter. One can sometimes walk about there without a hat, coat, or gloves. Some outer parts of the continent have summers that are much like winters in central North America—quite livable.

Antarctica's visitors see penguins, seals, water birds, and whales, and icebergs, glaciers, and snow, scientists at work, the Ross Ice Shelf, the sky lights of the *aurora australis*, smoking Mount Erebus, and 10,000-foot mountains rising from the water's edge. They also see the huts erected between 1901 and 1911 by the British explorers Scott and Shackleton. Caretakers now look after these historic buildings as there is concern for their preservation. They and their contents lasted in almost perfect condition through the cold and storms of all those years, but will not survive human visitation unless guarded. (They were ransacked before tourists began arriving.) They contain equipment and furnishings and still-edible food just as the early explorers left them when on their way to the South Pole, some never to return alive. New Zealanders set up a zoo-museum at Cape Royds and provide guides for tourist groups.

Ski-equipped planes will probably soon be taking tourists to interior points, including the South Pole. Mountaineers and skiers are eager to visit Antarctica. The continent is full of peaks that have never been climbed, or even named. Mountain climbers have already flown in and attacked the highest peaks, over 16,000 feet tall. And who would not love to ski Antarctica—the ultimate!—where ski runs are always in top condition. In 1966 Antarctica already had a ski tow, strung up by New Zealanders 840 miles from the South Pole. The rope tow was driven by wheels of a tracked vehicle mounted on oil drums. Skiers in our part of the world often ski in weather more inclement than they would find at times in Antarctica.

Camping is coming to Antarctica, and small-plane owners are demanding access. Though 98 per cent of that continent is ice-covered, some peninsulas, valleys, islands, and part of the shoreline are ice-free.

Geologist Wakefield Dort, Jr., camped in Antarctica in a small tent and wrote of his experiences in the Sierra Club *Bulletin.* "This is a strange land," he said, ". . . alien and deadly to those who neglect to respect its ways." In spite of the cold, the blizzards, and the dreariness, he found: "The summer sun is welcomed, almost worshipped. Its radiant heat warms dark rocks that thus become wonderful to sit on or lean against, and it melts a little snow or ice to provide the only source of water for drinking or cooking. When the sun is shining brightly, its heat on one's back causes a vigorous sense of well-being; a thick cloud cover brings despair."

There was no vegetation except lichens, and no birds. "Deep breaths find none of the zestful aromas of spruce forest or sun-hot meadow, no flowers or spicy herbs, not even 'mountain freshness.' The air is sterile, odorless, blank." On the other hand, the scenery was magnificent, the vistas almost unbelievably wide and long in the clear air, and there was no need to worry about visits from hungry animals. The silence was complete. Many a modern-world citizen would gladly travel half the globe to find such silence and purity. Campers being what they are—hardy, nature-loving, adventuresome—they will be there.

Tourists are definitely invading the coldest and iciest regions on Earth. TV viewers have watched snowmobile treks over the frozen Arctic Ocean toward the North Pole. The Admiral Richard E. Byrd Polar Center aims to acquaint people with polar regions and has been doing so with tours to Arctic Ocean islands and flights over both poles. Various organizations are now sponsoring Antarctic tours. And people are waiting in line to go to the remote ice-locked corners of the world. Not all of these are in polar regions.

Mount Everest, the world's highest mountain, might be considered as secure as any place could be from tourist erosion, but the movement to it has already begun. Although many mountaineers have responded to its lure, their comings and goings have left the lofty, ice-draped bastion unscarred and their footprints were gone with the first wind that blew. But as tourists and parties of skiers converge on the mountain *en masse* they could leave more noticeable, lasting imprints. Mountain-climbing expeditions used to spend weeks plodding over rugged approaches just to reach the base of the 29,028-foot peak, but now an airfield has been built close to it. Other glaciered peaks and ranges of High Asia have been brought into easier striking distance too. Climbers book individual mountains in advance through the local government by obtaining permission and paying a fee. Most of the thirty peaks above 25,000 feet in the Himalayas have already been climbed, but about a hundred above 20,000 are still waiting their first conquerors.

Almighty Everest is being closed in on. One now finds a strange building on the mountain's flank—a hotel, a joint venture of Nepal and Japanese investors. Lack of oxygen is always cited as one reason more people cannot go to the mountains, but that obstacle is being eliminated. Guests arrive by plane and then are brought to the hotel by pony-drawn carriage. An oxygen mask is supplied on landing. Because guests of the hotel at 13,000 feet may need time to acclimatize to the rarefied atmosphere, beds are covered with oxygen tents. A small hospital is close by. The hotel, which is managed by the Japanese, was built in spite of great expense and tremendous construction problems. Sherpa tribesmen, who are the porters for climbing expeditions, staff the hotel.

If Abominable Snowmen are in those Asian mountains they are going to have company. And if Shangri-la exists there it will be found.

Tourist erosion is not to be considered all bad necessarily. It is as inevitable as natural erosion. Tourists cannot be kept away from glaciers, wherever they may be, nor would we want them to be, especially since we are all tourists ourselves. It is one of the rewards of progress that those of us who do not have the physique, stamina, and nerve of mountaineers can now have mountaintop experiences too.

We who are infirm or fearful or old need no longer stand at the base of glaciered peaks looking up as mere human beings, but can stand at the fountainhead looking over the world below with the perspective of a god—or a glacier. We can see the whiteness pouring down and feel the dizzy elation that heretofore was reserved for the hardy few.

However, one who is deposited on a mountain glacier by helicopter or ski plane or cablecar minutes after having left warmth and safety cannot appreciate the view and exhilaration or feel the climactic emotion of being there as can those who have scrambled up powered by their own strength and drive.

John Tyndall, the noted British naturalist and mountaineer of the last century, was one who knew that transcendental feeling. He would often go up into the icy mountains alone, a danger for even an experienced climber such as he. In his book *The Glaciers of the Alps and Mountaineering in 1861*, he wrote, with the verve of scientists of old:

I was soon upon the ice, once more alone, as I delight to be at times. . . . When your work is clearly within your power, when long practice has enabled you to trust your own eye and judgment in unravelling crevasses, and your own axe and arm in subduing their more serious difficulties, it is an entirely new experience to be alone amid those sublime scenes. The peaks wear a more solemn aspect, the sun shines with a more effectual fire, the blue of heaven is more deep and awful, the air seems instinct with religion, and the hard heart of man is made as tender as a child's. In places where the danger is not too great, but where a certain amount of skill and energy are required, the feeling of self-reliance is inexpressibly sweet, and you contract a closer friendship with the universe in virtue of your more intimate contact with its parts. . . .

But time is advancing, and I am growing old; over my left ear, and here and there amid my whiskers, the grey hairs are beginning to peep out. Some few years hence, when the stiffness which belongs to age has unfitted me for anything better, chamois hunting or the Scotch Highlands may suffice; but for the present let me breathe the air of the highest Alps.

Were Tyndall here today he could be lifted easily to those heights in a matter of minutes.

Lucky we are to be living before the effects of overpopulation

reach all the unspoiled parts of the world, while we can still view a
glacier without being jostled by elbows, still hear the nothing-silence
that envelopes it except for the wind and the glacier's self-made
sounds, still watch in the evening solitude as alpenglow plays on
white mountain peaks, turning them shades of pink with the last
tangent rays of the already-vanished sun.

Eventually tourists will come in droves and glaciers will show the
effect. More of them will lose their pristine quality. Their silver surface will bear the tarnish of litter, soil, foot tracks, and vehicle ruts.
They will melt faster, other things being equal, because of physical
wear and the imported warmth of civilization that is part of tourist
erosion.

We should not try to keep tourists from glaciers. We should want
everyone possible to see them and feel the thrill of being near them
or on them. Tourist erosion then is bound to occur. But it is hoped
it can be controlled so glaciers and their surroundings can retain as
much as possible of their natural character.

Glaciers now are fair game for anyone. They can be used as roads,
airports, ski runs, and playgrounds. They can be chopped up, hollowed out, and their habitat can be totally transformed, all for sport
and fun and profit.

Perhaps you will think as I did when you read Glacier National Park's descriptive folder which says, "Your visit to Glacier
National Park can be a rewarding experience. . . . Please help protect
this park and all it contains."

CHAPTER 23

OUR ICEWARD MIGRATION

The winter! the brightness that blinds you,
    The white land locked tight as a drum,
The cold fear that follows and finds you,
    The silence that bludgeons you dumb.
The snows that are older than history,
    The woods where the weird shadows slant;
The stillness, the moonlight, the mystery,
    I've bade 'em good-by—but I can't.

. . . .

There's gold, and it's haunting and haunting;
    It's luring me on as of old;
Yet it isn't the gold that I'm wanting
    So much as just finding the gold.
It's the great, big, broad land 'way up yonder,
    It's the forests where silence has lease;
It's the beauty that thrills me with wonder,
    It's the stillness that fills me with peace.*

ROBERT W. SERVICE

TACITUS, the Roman historian, believed that no one from the sunny Mediterranean shores would voluntarily live north of the Alps. Colder, backward northern Europe was considered virtually uninhabitable for "civilized" people at that time, about A.D. 100. And the offshore, foggy, rain-soaked British Isles? Quite out of the question. For all his learning he but spoke the general opinion of his day.

* From "The Spell of the Yukon" by Robert Service from *The Collected Poems of Robert Service*. Reprinted by permission of Dodd, Mead and Company, Inc., McGraw-Hill Ryerson, Ltd., and Ernest Benn Ltd.

But many of us, even with a longer ken of history, still hold the age-old belief that undeveloped regions colder than our own cannot possibly be comfortable, healthful, prosperous places to live in and so will never support large permanent populations. We forget that people since their origin have been migrating toward the cold. The always-warm regions seem to have been their birthplace but they did not hold them. Apparently Early People originated as tropical or subtropical creatures, but after they had tamed fire many became well established in harsher climates of the middle latitudes. They made the gradual, unconscious choice to live there. They had reasoning far above any animal's and even animals migrate to areas they prefer. Once Early People had the means to survive in wintry lands they occupied them, though gathering food and protecting themselves from the elements required more effort there than in always-warm realms.

Early People were in the middle latitudes of Eurasia since the third-last glacial stage at least, but we do not know how the populations were distributed. During the final glacial stage, hunting bands were in the Far North in Siberia and crossing over to Alaska. And as soon as the last ice sheets in North America and Europe melted back, people of our species roamed into the vacated territory. Long before all the ice lumps had melted from the ground, before the bogs had drained, before the trees grew back—our kind of people were there. (This should be no surprise, for many people today choose to live in environments colder than that.) By then they not only possessed fire but constructed primitive shelters and sewed skins together to make clothes. Nothing short of land's end or remnant ice caps halted their coldward migration.

With the birth of agriculture settled living and denser populations became possible. People began clustering together in sedentary communities in the Middle East and probably elsewhere about nine thousand years ago, while the last fragments of the ice sheets disintegrated over Scandinavia and northeastern North America. Irrigation, animal husbandry, use of metals, and other cultural advances helped populations grow and furthered control of the environment; and new waves of migration rippled across the continents from various centers. Not migrations of nomadic hunting bands alone, but also of creative ideas, of field cultivation, of community living.

At the dawn of historic times we discern a particular wave that crested in the North Africa–Middle East region and slowly rolled northwest toward where the ice sheet had been. This wave was the center of civilization and power. This phenomenon is sometimes used to argue that cold, moist, stormy climate is not detrimental to mankind, but stimulating and beneficial. (The wave of civilization spread in other directions too, but it is in this area that we have the clearest record of it. There the northwestward flow of post-glacial culture can be traced through historic time.)

While primitive hunters and experimental farmers occupied the North, civilization was rising to sparkling heights in Ancient Egypt and Mesopotamia. Other parts of the Fertile Crescent and the eastern Mediterranean area were included in the effervescent tide of superior culture which developed irrigation, community organization, writing, religion, art, and architecture. The influence of these areas ebbed, and focal points of leadership and creative vigor ultimately shifted away from the North Africa–Middle East–Asia Minor area to the north shore of the Mediterranean, and there emerged "the glory that was Greece" followed by "the grandeur that was Rome" (Poe). Mediterranean countries led for a long time in art, science, exploration, and power; but then the wave of creative energy splashed north of the Alps to the region Tacitus scorned, and northern European countries surged to dominance in modern history.

Ellsworth Huntington called attention to the way the center of civilization migrated northward in *Civilization and Climate* (1915). He believed that when the early seats of superior culture were enjoying their periods of greatness their climate was different, more like that which northern Europe, the United States, and southern Canada experience today—cooler, stormier, more humid. This was because they were then more strongly influenced by the middle-latitude storm tracks, the convergence zone where cool air masses from the north mix with warmer air masses from the south. This belt of changeable weather with continuously passing "highs" and "lows" moves north and south in response to atmospheric pressure patterns. Theoretically it was far enough south to have had an increased effect on the Egypt and Middle East area several thousand years ago and then it moved north. While the ice sheets covered the North, they produced strong high atmospheric pressure and outward-pouring cold air masses. Therefore the storm tracks were kept farther south. As the Ice Age waned gradually and with fluctuations, the high-pressure conditions weakened in the North and the storm belt could move north. (A latitudinal shift of storm tracks occurs yearly now. In winter the tracks follow the sun south as northern air masses exert strong out-flowing force, and then the Mediterranean Basin has its rainy season. In summer, when cold air masses weaken, the convergence belt moves back north, and desert air masses from North Africa move north too, giving the Mediterranean Basin its dry season.)

While the various influential nations around the Mediterranean were at their peaks of creativity and vigor, their climate supposedly was not as dry as now, and storms passed with greater frequency, giving energizing variety to the weather. Perhaps then, those thousands of years ago, Memphis, Egypt, had weather more like Memphis, Tennessee; and Babylon, Tyre, Athens, and Rome may have had weather more like Paris or Amsterdam or New York or San Francisco. The succession of storms; the cooler weather; the sudden, unpredictable fluctuations in temperature, pressure, humidity, and winds, as first one air mass enveloped the area and then a contrasting one; the variability and uncertainty of weather—these were stimuli that kept people alert and active, requiring them to work hard and quickly, and plan ahead. Higher precipitation also assured an ample food and water supply.

Those who adhere to this theory that storminess and cooler weather produce more vigorous, creative people, will point out that much of the tropics and subtropics is, in general, less productive agriculturally, industrially, and culturally than middle latitudes despite having a year-round growing and outdoor-living season. They point out that if present warm-to-hot regions are successful in commerce, industry, and agriculture they are so because of in-migration of people from colder regions who have injected their energy into the society. It is pointed out also that when people from middle latitudes move to the tropics they usually live in the highlands where it is cool, or do little physical work, or stay there only temporarily, or live in air-conditioned buildings, send their children north to school, and so on; but that if they remain there in the tropics a long while they or their descendants usually adopt a less active mode of life.

Of those who have explored this aspect of climatic determinism some strongly reject it, for the theory does have weaknesses, but others as strongly support it.

Arctic explorer and anthropologist Vilhjalmur Stefansson, in his book *The Northward Course of Empire*, said of Man, "His fight upward in civilization has coincided in part at least with his march northward over the earth into a cooler, clearer, more bracing air."

The march north continues, and the world's population growth is only one reason for it. People could as easily emigrate to sparsely populated areas in warm latitudes, such as the longed-for desert isles, or the Amazon and Congo basins, the savannas and arid lands. And some do. But there are indications that the promise of future rewards is not as great in warm regions as toward higher latitudes. A high percentage of the people from middle latitudes who "go south" even a short distance do so only during vacations or to retire, not to make their fortune.

It is popular opinion that if people were free to go where they chose they would all go to the tropics and subtropics, that these are the much-preferred climates. Then why don't more go there? Only a small fraction of people born in cold climates leave them, and some go into still colder regions. Large numbers of Negroes and other warm-climate ethnic groups have been moving poleward; one might say the reason is economic, and if so that is part of this total migration pattern.

Stefansson, who was Manitoba-born, tells how, when the Canadian Pacific Railway was being planned, a commission in England was to decide whether the Canadian West was suitable for settlement by British emigrants. After hearing witnesses and learning of the blizzards and temperatures 50° below zero, the commission concluded this was polar climate and that Manitoba and Saskatchewan were unsuitable for colonization by average Europeans. We smile at this now, but many people today have precisely the same notion about the next provinces to the north. The Northwest Territories, the Yukon, along with Alaska and Siberia, used to be considered iceboxes in which no sane person would live, and yet they are now inviting frontiers with rapid in-migration.

Perhaps the climate is warming in the North, though weather statistics do not yet confirm the fact. Perhaps many among us still respond to the ancient call of coolness and open space, unaware that they are participating in a mass migration that began in prehistoric time.

Still partly nomad, we keep moving, or our children do. The congenital hunting urge remains strong. Sportsmen fly far into the wilds of Canada and Alaska to fish and to hunt caribou, moose, big bear, and other animals. Modern resorts provide accommodations amid the primeval forests and tundra. Prospectors for minerals are also there, and the naturalists, engineers, and others who have helped open the once-empty lands.

Persons familiar with cold climates know they are not as unpleasant as rumor says they are. One must see for oneself. We shun and mistrust the unknown.

Stefansson found the Far North a completely hospitable place, much more to his liking than southern regions. In the above-mentioned book he wrote:

When we stop to analyze the expression "a good climate" we find that what we really mean is a good climate for loafing. Secondarily we may mean a climate where all sorts of vegetation flourish rankly. Without denying the value to the world of coffee and cotton and sugar, we are constrained to admit that the most important crop of any country is the people. No climate can rightly be considered good, though bananas and yams may flourish, if men decay. Human energy, mental and physical, is developed to the highest degree in the northern climates. It may also in some cases be developed to a high degree in southern countries, notably on plateaus and where the sea breezes blow freshly.

In another book, *The Friendly Arctic*, which describes five years of living with Eskimos, Stefansson says, "When the world was once known to be round there was no difficulty in finding many navigators to sail around it. When the polar regions are once understood to be friendly and fruitful, men will quickly and easily penetrate their deepest recesses."

Word has spread and the penetration is going on. The last realms of cold and ice are about to lose their long-perfect naturalness.

Those who have been to the Far North know it is not an icy, marshy waste fit just for reindeer and Laplanders, walruses and Eskimos. Explorers in the subarctic and Arctic thought they would find deep snows but precipitation was light. They thought food would be scarce but it was more plentiful than they expected. There were many large and small animals on land, countless birds, fish, seals, and other sea life. Agricultural experiment stations have developed plants that grow well in low-sun, long-day regions, and greenhouse agriculture is becoming important. The muskox has been domesticated, and is raised not only for meat and hides but for its silky underwool, which can be spun into a delicate yarn finer than cashmere. In its Arctic habitat this Ice Age animal needs no barn for winter protection (only protection from Man) or any feed other than its native tundra.

The Canadian government is strongly encouraging development of its colder regions—the untapped mineral wealth in the Precambrian Shield and the mountains; the oil and gas reserves in the northern prairies and along the Arctic coast; timber and water resources; spacious recreation country in snowy mountains, in shaggy taiga forests, and among glacier-formed, pure lakes. This is not idle country any longer. It is a rich frontier for prospectors, prime vacation land, and just plain good country in which to live.

Canada, the world's second-largest country (after the USSR), has been a sleeping giant but is fast waking up. As resources of earlier-developed countries are being depleted her rich reserves are eagerly sought. That wide country's surface has only been scratched. Claims are being hurriedly staked. Added highways and airstrips are making remote spots accessible. Railroad lines are being extended. Instant towns are springing up, and picturesque native villages are becoming modern. Yellowknife on Great Slave Lake now has paved streets, a golf course, and tennis courts, and it draws thousands of visitors annually. In the High Arctic, Inuvik on the Mackenzie River delta is a tourist stop, and at an Eskimo village on the Arctic Ocean coast tourists can sleep overnight in igloos built just for them.

Snow and ice have not thus far kept determined people from going into and living in any region they wish. At first not many wanted to venture into cold regions—a few hardy explorers, missionaries, traders, trappers, sourdoughs, whalers, geologists. Now that the scouting has been done, others have joined the coldward migration—miners, lumbermen, industrialists, sportsmen, sight-seers, homesteaders, businessmen—until finally the full movement is on to open the cold regions, subdue them, and dress them up for people with middle-latitude living standards.

This opening up of cold regions, even glacier regions, has happened in a shorter time and with less difficulty than was expected. Were the obstacles perhaps not as great as had been supposed?

Stefansson made this frank admission based on his own experience:

That is the beauty of being a polar explorer. You can go far away and do things that are easy to do, come back and say the country is friendly and the work pleasant, and still get credit for being a hero who must necessarily have gone through terrifying adventures in a region of utter desolation!

He is confirmed by Paul-Émile Victor, who organized many expeditions to both polar regions. He wrote in *Man and the Conquest of the Poles*:

We must point out a slight failing in the polar explorers: they did not hesitate, in their accounts, to depict good-naturedly their anguish and their sufferings, which were sure to give them something of the look of heroes and which often moved those who had sent them to raise their salaries. This cult of heroism has been long kept up by the explorers, and by some of the greatest. . . .

The modern explorer . . . goes because it pleases him to work in the polar regions rather than elsewhere. He does not feel that he will become a hero simply for this reason. He does not expect to suffer, and, as a result, he usually doesn't.

But during the period that we call heroic—which, in many ways, it was —it was in certain men's interest to be taken for heroes and to preserve the hellish reputation of the polar regions; men with the call turned up all the more rarely, and the fishermen as well as the fur traders could get on with their business in the midst of the ice floes.

Since we are migrating iceward it helps to know that our bodies become acclimatized to colder weather. Weather that seems uncomfortably cold at first seems less cold after a period of exposure. Animal experiments and human experiences show acclimatization is possible.

Workers in polar areas have shown adaptation in fingers and hands. Eskimo babies play nude in below-freezing temperatures on the floors of igloos. Australian aborigines slept naked on the ground in near-freezing temperatures.

When Charles Darwin voyaged around the southern tip of South America in 1832, he was shocked to see that Indians there wore little or no clothing despite the windy, cloudy, cold, "wretched" weather typical of that Tierra del Fuego area. There was snow in the hills and sleet in the valleys. Men and women without clothes canoed in heavy rain and dove into the ocean for seafood. A mother nursed a naked baby while sleet fell on her bare breast.

When Europeans came to America they expressed surprise that the Indians went half naked in cold weather while they, fully clothed, were shivering. The Indians' explanation, the story goes, was: "White man's face not cold. Indian 'face' all over."

When Admiral Peary reached the North Pole his Negro manservant was with him.

Acclimatization does seem to work.

The movement into cold regions is accelerating, and one of the strongest deterrents, the transportation problem, is being overcome.

"One might say that the snow as an impediment to winter travel has been beaten," reported a Canadian committee on soil and snow mechanics. Canada is a leader in snow removal and ice-control techniques. In cities snow is combated not only by hauling it away and by using chemicals; it is also melted in burners along streets and some pavement heating is done. There is actually far heavier snow in southern Canada than in northern, so if the South can master the snow problem the North will do so too.

Hovercraft (vehicles that ride on a cushion of air) skim over open water and sea ice without touching the surface. In Alaska, railroad tracks have sometimes been laid right on a glacier; they have to be straightened when the ice moves, of course. In severe cold, plug-in electric cables keep engines of parked cars from freezing.

Canadians and Americans are extending the St. Lawrence Seaway shipping season, which in the northern part of the Great Lakes normally runs from mid-April to mid-December. Navigation channels are kept open by using icebreakers, dynamite, and infrared radiation, by piping steam under the ice, and by pumping bubbles through the water. In higher latitudes ice on waterways is dusted with dark particles and hosed with salty sea water to induce melting.

The USSR uses icebreakers in the Arctic Sea route along the Siberian coast and has been fairly successful in lengthening the shipping season there. Dusting of the ice has helped too, advancing ice break-up in Arctic bays as much as a month and opening leads even in Vilkitski Strait around Eurasia's northernmost headland. This

Northeast Passage is commercially and strategically important to the USSR because it connects the Atlantic and Pacific; because it follows her longest coast; and because Siberia's main rivers, which are highways of the interior, flow north to the Arctic Ocean. Siberia is rapidly being developed because of its vast store of resources, including uranium, diamonds, coal, gas, oil, hydroelectric power, forests, gold, tin, nickel, platinum, iron, and so on.

Across the pole North Americans are attacking the sea ice cover from the other side. Icebreakers clear lanes along the shores and among the islands, and in 1969 United States submarines traveled under the polar ice, pioneering what may become a commercial cargo route under water, under ice, through the Arctic Ocean.

The slogan for Alaska's 1967 centennial was appropriately "North to the Future." That state, like Canada and Siberia, is a rich cache of coveted resources, and its latest spur to development is the oil boom on the Arctic coast. The oil of Alaska's northern coastal plain and sea floor is believed to be a reserve equal to that of the Texas fields, and the continuation of that field along Canada's north coast may equal that of the Middle East. When an icebreaking oil tanker rammed through the formidable frozen Northwest Passage from the Atlantic to northern Alaska, proving commercial shipping was possible, it was said that it opened the Arctic not only physically but in people's minds.

Technology stands at the edge of the ice, and the ice will not block it much longer.

Japanese shippers are studying possible sea routes to Europe via the Arctic Ocean, which would shorten the trip from three months to one. They think such a route through and around the ice is feasible and that conditions are not as restrictive as generally believed. Transpolar shipping would, of course, be advantageous for other countries as well.

The Arctic Ocean ice cover may be an unstable thing. If broken enough it might disappear altogether and permanently. When ships cut through the ice pack, making open sea lanes, heat from the ocean escapes to warm the air. Smoke and dirt darken the ice, causing warming and melting. The Arctic ice cover is not protected by treaties or agreements. Economic pressures or military requirements could conceivably lead to full-scale efforts to open the ocean.

North America's Arctic shore ceases to be the Eskimos' domain as industry and commerce come in and hot oil gushes up from Alaskan and Canadian wells. Spillage from wells, pipelines, and tankers will be inevitable. When a snowslide recently tore through a tank farm on a northern Quebec shore, diesel fuel and gasoline poured loose, and it was reported that the best way found to get rid of the spill was to burn it as it floated on the water or to pump it onto sea ice and burn it there.

So as the Arctic coastal plain, islands, and sea floor are pierced by oil wells we can foresee the whiteness turn black with oil and the dirt of drilling and shipping, and even see it aflame. Hardly the familiar picture of unspotted ice floes and polar bears.

Mechanization can destroy the firmness of the tundra. Even the track of a light vehicle driven once over the land is enough to ruin the protective vegetation and start the ground slumping. The tracks enlarge and deepen to ditches, scars which may last centuries in the slow-healing tundra.

As our remaining Ice Age environments are found to be not worthless but commercially valuable, livable places, they are bound to suffer the defacement that usually comes with exploitation and development. *Science* reported an interview with the director of an environmental laboratory of the University of Alaska a few years ago when conservationists were first challenging the planned oil pipeline across Alaska, and the director reportedly said, "The tundra is delicate, but so what? If it's torn up, esthetically that's poor. But if there is nobody there to look at it, what's the difference?"

Many Alaskans—fortunately not all—would willingly sacrifice natural, undefiled landscape for another Gold Rush. That was the biggest event that ever happened there and they would like to see a repeat. There will be more rushes, for other commodities if not for gold. But they may not realize that one resource more valuable and scarce than gold *is* natural, undefiled landscape. This they still have, and quietude large enough for a person to get lost in. These vanishing treasures cannot be recycled or stored in warehouses and vaults; they have no substitutes; and they are vital to mankind's sanity and health. They are already being rationed in more populated areas as admissions to parks and natural areas are being limited, and they are being hoarded, as individuals and governments rush to buy land in its natural state before it is all gone.

The long-inhabited areas wear out and turn shabby. The sedentary population stagnates from lack of energy outlets. So the hardy, the ambitious, the daring, those repelled by the suffocation of stifling living look to a new invigorating territory where they can unleash their energies to the full. Outer space may be the horizon of the more distant future, but for now the cold regions offer the challenge and opportunity.

Greenland is not the desolate island it once was. Since recolonization began in 1721 its population has grown to about 50,000. Today one sees mines, factories, farms, supermarkets, apartment dwellings, and military and science research bases. Its main ports operate all year. Fishing is the leading occupation, having supplanted seal hunting in importance early in the century when seals diminished in numbers and were replaced by fish, a sign of warming climate. Cod, shrimp, and salmon are processed in modern factories on Greenland's west coast.

Canada's Baffin Island is coming to life. Treeless and barren, wearing large ice caps, it used to be thought of as a nothing-land, but now it looks attractive. Its extremely high-grade iron ore will not remain unmined for long. A few years ago mountaineers explored its spectacular mountains and found climbing could be done leisurely there north of the Arctic Circle with twenty-four hours of daylight in summer. The weather was unexpectedly comfortable. They found the July temperature rarely fell below freezing and often reached into the seventies. Many days were warm and windless with bright sunlight. Now major climbing expeditions are planned there. Access to Baffin Island is no longer difficult, as airlines have regular flights to most settlements, including jet service to Frobisher Bay, the southern harbor which Canada is actively developing. One going to that little-known island might be surprised to find such events as spring proms, ball games, and fireworks on New Year's Eve.

Fireworks light the skies of Antarctica too. When that mightiest citadel of crystal cold fails to halt mankind's migration can anything real be left of the Ice Age?

Its ice-packed ocean moat which used to repel landing craft is no longer totally effective. Icebreakers open channels and freighters dock at its coast. Planes play leapfrog over every obstacle.

Less than a half century ago, when Rear Admiral Richard E. Byrd and his 1928–30 party were left there to winter over, he said, "When the winter night should set in, all the combined merchant marines and navies of the world could not reach us." And in 1934, when he bravely spent the winter there alone, he nearly died from stove fumes in his shack before being rescued the next summer. Now even in the black of winter planes fly in. There is no section of the continent in which a base could not be comfortably maintained year-round.

Within a few decades after Byrd's expeditions the situation had altered almost beyond belief. Another American rear admiral, David M. Tyree, who headed Operation Deep Freeze from 1959 to 1962, said, "We are in Antarctica to stay." McMurdo, America's largest base there, is now a good-sized village boasting a nuclear power plant, a water-desalting plant, a fire station, street lights, tourists, a bronze bust of Admiral Byrd on a marble pedestal, warehouses, and modern buildings with offices, lounges, shops, and recreation facilities. Several hundred people winter over at that one base alone and a couple of thousand live there in summer; and additional smaller bases of the United States and other countries dot the continent. The United States' new South Pole station is covered with a geodesic dome. The total winter population of Antarctica is about 500 now and it will be increasing.

Going to Antarctica nowadays is not the danger it was in the past. In recounting his Antarctic experiences to the Royal Canadian Geographical Society, O. M. Solandt, a past president of the society, said:

I would strongly recommend that any of you who are invited to go to the Antarctic accept the invitation. It is a perfectly fascinating experience and not physically demanding. The climate is certainly better than our winter climate in Ottawa. The food is excellent, the transportation is good, and, above all, it's an extremely peaceful place.

The first woman stepped ashore in Antarctica in 1935—a Norwegian ship captain's wife. Since then many women have come there. First just privileged wives, then tourists, and now women scientists as well.

Twelve nations signed a thirty-year treaty, effective until 1990, in which they agree to use Antarctica for peaceful purposes only, to allow the other countries to inspect their operations, and to preserve the continent's resources and environment. A number of these countries had previously laid claim to parts of the continent and some claims overlap, a source of future trouble perhaps. (The United States has made no claim though it conducts the largest operation there; on the other hand, it does not recognize other countries' claims.) Geologists are searching out mineral deposits and it is uncertain what will happen when the treaty is no longer in effect.

Antarctica is still the cleanest place on Earth. Bacteria do poorly there and the intense cold is a preservative. Meats, dried fruit, biscuits, and fodder left by Scott's and Shackleton's expeditions were well preserved after being there more than fifty years. Far from large industrial and populated areas, it undoubtedly receives less pollution than any part of the world too, although fallout it has received lately includes radioactive material from nuclear explosions, chemicals from cloud-seeding operations, and lead from Northern Hemisphere industries. Efforts are made to minimize contamination and environment disturbance but the results are unsatisfactory. DDT has been found in penguins and sea life. Penguins are bothered by aircraft and onlookers, and evicted from nesting grounds; and stray sled dogs have grown fat eating them. Oil from ships coats the water. Broken-down vehicles and planes are commonly left where they are stranded. Smoke causes camp smogs. Untreated sewage emanates from many camps. There is no subsurface drainage and bacterial decomposition is extremely slow. Garbage and other polluting refuse are thrown into the sea. Biologists doing ecological studies have to go continually farther from camps to find natural conditions.

It is being suggested that Antarctica be used as a dump for radioactive waste which would be flown there in sealed containers and dropped onto the continent's interior. The slowly creeping ice would bring them to the ocean after several thousand years—if it did not crunch and break them on the way. The ice sheet is also eyed as a cold-storage warehouse for food and other perishable products. Even for human bodies. Some people who believe medical science may in future years know how to cure their now-incurable affliction, or rejuvenate them, are having themselves frozen when they die, with the plan that they will be thawed years later, be repaired, and go on living. Cold-storage vaults at home will preserve them as long as the electricity does not go off, but they might be safer longer in a nearly stationary part of the Antarctic ice sheet. Close check would be made to see that they were undisturbed as the ice slowly moved. If a catastrophe wiped out their homeland they would presumably remain

safe there. It is eerie to ponder what would happen if no one were
around to revive them when the time came, before, some distant
day, they sailed forth in an iceberg or wound up in a moraine.

Antarctica is definitely open and occupied and proving to be an
invaluable piece of real estate in this scientific age. It is strategically
located for transpolar flights, is the only land area at a pole, and as
such is an ideal, open-air laboratory for biologists, climatologists,
physicists, space scientists, astronomers, and other researchers. It has
a naturally near-sterile environment with simple, closed ecological
communities. It contains the south geographic and geomagnetic
poles. It has the coldest temperatures and strongest winds. Its long
winter midnight and black-as-can-be skies make it an excellent observatory, and its quietness is superb for radio monitoring. It is a
powerful affecter of weather and ocean circulation. Geologically it is
the missing link in the story of Earth's formation.

So we are migrating poleward, as we have been since the ice sheets
withdrew, toward colder and colder regions, to the very ice itself. In
lands once thought too cold to be habitable farms are prospering,
mineral resources are attracting industry and investors, and villages
have mushroomed into modern cities. Tourists and sportsmen are
flying to glaciered mountains and to the taiga and tundra with relative ease. The Arctic Ocean is encircled by would-be conquerors
camped around its shores. Its cracked and drifting ice cover is being
riven and sailed under, and ice floes are used as floating research stations and airports. Even in Greenland and Antarctica people are living in comfort and security right next to the last ice sheets—in fact,
right *on* them, and *in* them!

We draw near to the ice in growing numbers and with growing con-
fidence. Ironically, the closer we come to the ice, the more the image
of the Ice Age fades.

CHAPTER 24

CIVILIZATION VERSUS THE ICE

Now the polar stage has been taken over by scientists and technicians, whose painstaking work really achieves much more than the flamboyant exploits of the earlier explorers. But the glamour is gone.

FRED BRUEMMER

WE HAVE lost our fear of glacial realms. When fear goes reverence goes, and when reverence goes desecration begins. The ice stands in the way of our conquest of our planet. It is just one more resistant obstacle to be overcome.

Probably no one has felt any compunctions about chopping away at glaciers or making them melt faster when they had the need or desire to do so. We have seen how glaciers are eroded by those who visit them and work among them, and how they are darkened with dust and dirt to speed their melting and increase streamflow. The deeds have not been questioned because they are either harmless when done to healthy glaciers, or far removed from the notice of environment-conscious societies, or done under the aegis of presumedly careful, well-intentioned groups like universities or government agencies.

Up to now the effects of ablative techniques used against glaciers, as well as sea ice, have been slight in most cases. Some projects are experimental, and the need to experiment seems to justify anything. One wonders what will happen as experiments become larger, as with climate modification; or when more powerful methods of destruction are used; or as certain segments of society find cause to do serious battle with individual glaciers or glaciers as a whole. If glaciers are assaulted in earnest they may be no more resistant than were forests against the ax or mammoths against the stone-tipped spear. The first good reason people had for removing ice from glaciers was to cool something somewhere else. Before artificial refrigeration came into use glacier ice was in demand for cooling perishable goods that were stored or shipped. Though it was more difficult to obtain and handle than ice cut from lakes in winter, it was colder and longer-lasting. Ships of the Alaska Ice Company used to carry ice to San Francisco, the Sandwich Islands (Hawaii), China, and Japan; and ice from Norwegian glaciers was exported to Great Britain and other places. Shippers of perishable goods still use glacier ice when feasible to refrigerate their carriers.

Now and then a glacier may be partially removed so that what is under it can be seen. For example, under a 14,000-foot-high glacier on Mount Ararat in eastern Turkey is the partly visible, collapsed remains of a massive wooden structure. Archeologists and theologians who think it might be Noah's ark want to look at the whole thing. Airplane pilots report that a ship-like shadow is seen in the ice when it is clear of surface snow. The Arctic Institute of North America agreed to help remove the part of the glacier covering the wooden object.

In the Alps much dynamiting is done in connection with road building and maintenance which shakes ice loose and increases melting. Glaciers there are dug into to make underground chambers for secret military redoubts, hideouts, and the storing of munitions and machines.

There are logical reasons why people might want to get rid of part or all of a glacier, and as our destructive powers grow, selected glaciers undoubtedly will be marked for reduction.

A "galloping" glacier might be one such target. Historical records tell how in the 1600's, during the Little Ice Age, rapidly advancing glaciers in the French Alps surged down their valleys into villages and piled up against houses, and the villagers in desperation chopped chunks of ice from the glaciers and hauled them away. Today if glaciers came grinding toward towns stronger measures would be taken to try to stop them. Before people are willing to sacrifice lifetime investments they will demand emergency action. Because people are settling closer to glaciers it is only a matter of time before the question of how to deal with a galloping glacier or even a normal growing one will require a practical answer.

Certain glaciers are looked upon as a public menace because they produce ruinous floods. They do this by creating lakes and then letting them discharge suddenly in a mass outpouring of water and ice. Such a glacier will extend itself, block a valley, and dam drainage behind it. A lake forms behind the glacier dam. Sometime later when the glacier shrinks, or when the pressure of the impounded water becomes sufficiently strong, or when the overflow begins to wear away the ice, the dam breaks; and with little or no warning the lake empties like a broken reservoir. In the lower valley bridges, roads, railroads, and power lines are destroyed, villages and farms are washed out, and there may be great loss of property and lives. Some ice-dammed lakes discharge like that every year, others at irregular intervals. Occasionally explosives are used on the glacier front to keep the ice dam from forming. Because this phenomenon occurs in many populated mountain areas, including the Andes, Alaska, central Asia, and Iceland, control of some of these threatening glaciers may be called for.

Other dangerous glaciers are those on or near volcanoes. They melt suddenly when hot molten magma nears the surface or the volcano erupts. Or they gradually melt at their bases from underground heat and collect subglacial reservoirs of water that burst loose unexpectedly. In Iceland, where glaciers overlie active volcanoes, a "glacier burst," or *jökulhlaup*, is often a shattering paroxysm, with floods of millions of tons of water, carrying volcanic debris, mud, rocks, and ice blocks as big as three-story buildings.

If public opinion and pressure groups were insistent enough, if the interests and welfare of large numbers of people were at stake, action might be contemplated against some threatening glaciers. A community might demand that a glacier be reduced or—if possible—removed to prevent recurring catastrophes, or perhaps to have a warmer local climate or space for expansion, or to clear a mountain pass for a highway, or to make a coastal inlet available for a harbor.

On a broader scale there might in due time be a concerted effort to remove some glacier ice for another reason: a scarcity of earth resources.

With world population multiplying at a frightening rate there has been alarm about the worsening food famine for some time, and lately it is realized that a fresh-water crisis exists too, but there is another shortage whose seriousness has not yet been grasped by the public—the shortage of minerals. We hear about the need to recycle metals, the stockpiling of critical materials, the depletion of certain deposits, but are not fully alerted to the fact that we may someday run out of essential minerals and that our way of life could collapse without them. Spokesmen of the U. S. Geological Survey have issued repeated warnings that our industrial civilization will begin to crumble within a few decades unless we can rapidly accelerate our ability to find and effectively use key natural resources. This has been considered exaggeration. But already the shortage of mineral fuels has upset our lives. Even before the 1973 cutback in Middle East oil exports, the fuel shortage was causing energy crises in transportation, farming, industry, and heating and cooling of buildings. In some places rationing had already become necessary or, worse still, fuel was just not available.

We have always ignored the possibility of a mineral famine because geologists have been efficiently scrambling about, finding new reserves fast enough to keep pace with consumption, but it is progressively harder to locate new reserves. We are down to using lower-grade ores. We are mining to dangerous depths, with some mines extending more than two miles below ground, to where the heat from the earth's interior is so intense that refrigeration must be used and even then the temperature is almost intolerable. This is the case in African gold mines. We are using substitute materials for metals, but for some metals there is no substitute. These measures cannot forever solve the mineral shortage. Already we are scraping shallow ocean floors and drilling into them; using sophisticated remote-sensing techniques to find new deposits; seeking ways to extract minerals from sea water; and hoping to find scarce minerals on the moon.

The visible part of the world's terrain has been fairly well examined by geologists. But 10 per cent is not visible, that part covered by glaciers.

Even if the world's population remained constant the prospect would be dire, but it keeps increasing at a phenomenal rate, particularly in Asia, Africa, and South America. An often-suggested remedy for the population explosion is to help the underdeveloped agricultural nations become industrialized so they can support a denser population and have a higher living standard. As underdeveloped countries are led into industrialization their new industries are going to need raw materials.

The United States, now with 6 per cent of the world population, is said to consume nearly one fourth of the world production of non-renewable resources. Its consumption of metals is increasing at approximately double the rate of its population growth.

Industrial nations of the world together, with about 20 per cent of the population, use at least half the world's mineral production, much of which comes from underdeveloped countries. It does not take higher mathematics to figure out that as the living standard rises in the underdeveloped countries the living standard in the United States, Europe, and other now-industrialized regions must decline, unless resources remain plentiful.

Minerals and mineral fuels are the basis of our modern industrial civilization. They are needed for energy, machines, construction, transportation, communication systems, fertilizers, household goods, chemical products, and so on. During the last few decades the consumption of these materials has exceeded the total consumed during all previous history! The demand for iron ore tripled in about twenty years. Mercury and tin have already become scarce. According to statistics gathered by the United Nations, known reserves of many minerals are adequate to meet anticipated demands for only a decade or two.

What will happen as underdeveloped countries clamor or are prodded to develop industries, to raise their living standards, expand building programs, have appliances, cars, planes, facilities, and luxuries that only a fraction of the earth's people have had till now? In some cases they will halt or restrict the export of their own minerals, which they are now glad to sell, because they will be using them themselves. They will compete for supplies of other countries. World demand for minerals will soar. The present affluent countries will have to be satisfied with less and their living standard may have to drop.

Geologists and oil and mining companies are often looked upon as enemies of the land by some conservationists and nature lovers. They are blamed for ruining the landscape as they extract resources, but we are the ones who are demanding the resources. As the geologists say, "Even the fisherman needs metal."

Our quest for minerals has given impetus to the iceward migration, for our search has brought us to the glacier-covered mountains, the polar seas, the ice sheets. Mineral resources of the cold regions have been among the last to be found and used because of problems of frozen ground, dark winters, transportation, and labor recruitment in remote areas; but now we must bring mineral resources from those cold regions.

Rich mineral deposits are probably hidden under glaciers and ice sheets, undisturbed as long as their presence is unknown. If we knew where they were we might do whatever would be necessary to get them. It is as though the ice were guardian, keeping these treasures out of sight and reach, saving them for future use.

Mining is already going on at the edge of glaciers and under them.

If a glacier hampers mining operations or stands in the way of reaching wanted minerals, what can or will be done? This situation has arisen at a number of places, some of which are in British Columbia. Near Smithers a tunnel was dug under a glacier to mine molybdenum. At another site a glacier snout is moving toward a mine portal and should bury it by the mid-1990's. Near Steward an eleven-mile tunnel was bored below large valley glaciers, which are thousands of feet across and as much as several thousand feet thick, to reach copper. Underground workings broke through the mine roof under the Leduc Glacier and subglacial water created a serious drainage problem.

Since special care must be taken when mining under a glacier— not to break through the roof rock to the ice because of the water under the glacier—it is natural for miners to think how much simpler it would be if they could reach the ore body from the top by strip mining. A mine spokesman told me that when such problems arise "consideration could be given to destruction or removal of the glacier."

When glaciers interfered with mining in years past, nothing could be done about it. Retreat of the San Francisco Glacier in Peru revealed ancient mine workings and tunnels begun by Spaniards. Near Mont Blanc is the French village of Argentière, named for a nearby silver mine which was worked during the Middle Ages but is now buried beneath a glacier. Nowadays, with the strong demand for mineral ores, miners might not give up so easily.

A large mining company has mined copper under a glacier in Alaska and more exploration is going on there now. The same company wants to begin open-pit mining in the Glacier Peak wilderness area in the northern Cascade Mountains northeast of Seattle. It owns the mineral rights but so far public pressure has delayed operations. Glacier Peak supports eight glaciers on its flanks and this is one of the most beautiful, truly alpine regions of the "lower 48" states. Objectors to the mine say it is not only the mine that would be unsightly, but with it would come a long access road, a mill, a tailing dump, settling pond, townsite, and a dusty, smoky atmosphere.

Gold is known to lie under certain Alaskan glaciers, and there are some gold seekers who will do just about anything to get gold. Ice on the Bering Sea has not stopped them from punching holes through the ice to reach gold in the sands below.

A claim for cinnabar (which contains mercury) and stibnite (which contains antimony) has been staked directly beneath Mount McKinley's picturesque north face, the Wickersham Wall, nearly in the geographic center of Mount McKinley National Park, and preliminary work has begun.

Glacier Bay National Monument, Alaska, is known to contain significant deposits of nickel, copper, and molybdenum as well as gold, silver, titanium, iron, and small amounts of coal and petroleum. The area was, unfortunately, proclaimed open to mining in 1936, and it still is, although to date mining has been negligible. The U. S. Geological Survey has made a reconnaissance appraisal of its mineral resources, and the thought of mines and ore-processing plants in that peaceful bay is a nightmare to park personnel and conservationists. The advisory board on national parks recommended that Glacier Bay be elevated to national-park status and that mining be banned, but Alaska's congressional delegation did not want mining forbidden, and park status may therefore be a long way off. The molybdenum, though low-grade, may be commercially minable soon because of price trends. Some of the nickel-copper deposits in the Brady Glacier area lie under hundreds of feet of ice, but these too may soon be minable. Brady Glacier (formerly Taylor Glacier) was the locale of the poignant, danger-filled story Stickeen, in which John Muir related his adventure with a lovable dog companion—recommended reading for anyone who loves dogs or glaciers or John Muir's writing.

Though little of Greenland's bedrock is exposed, mining and quarrying have been going on there. For years Greenland was the only source of the rare mineral cryolite, but that deposit is now exhausted. Beds of low-grade coal crop out there, as well as graphite, copper, molybdenum, lead and zinc, nickel, and radioactive minerals.

Antarctica undoubtedly contains considerable unknown mineral wealth. It may contain oil. Parts of other continents to which it once was apparently connected are mineral-rich. Eastern Antarctica seems to be a continuation of South Africa geologically and should it happen to have the same wealth of diamonds could glaciers covering them remain undisturbed?

As world population multiplies, as more countries become industrialized, as living standards rise worldwide, the press for under-glacier mining will increase. War emergencies could make it absolutely necessary for some countries. Until now there has been no practical way to handle a job as large as getting rid of a glacier or breaking through one, but maybe now the means and power are becoming available. Anyhow, thoughts are running in that direction.

Solar furnaces can concentrate the sun's heat intensely in a given spot. Space satellites will soon be able to beam concentrated solar energy to any desired point on Earth. Russian engineers have proposed putting artificial dust clouds at high altitudes to harness the sun's heat and would position them in such a way as to melt ice in cold regions.

It has been proposed that powerful nuclear explosives be used to blast away rock to create canals, new harbors, and passes through mountains, and to make mining easier by removing heavy overburden. That overburden in some cases could be a glacier.

A leading oil company has published ads picturing a glacier and stating that the company supplies enough energy to melt 7 million tons of glacier a day. From a practical standpoint, the company could not do that, nor had it any intention of doing so, of course, but the boast has an ominous tone.

That some communities might see advantages in reducing glaciers in their locality is understandable, and there is not much that glacier savers could do about the matter. Not all people or nations have the same scale of values. To us leopards are beautiful, rare beasts on the verge of extinction, needing protection; but to people living near them they are killers of livestock, a personal danger, and extra spending money if their skins can be sold.

A nationally prominent scientist told me, in jest of course, "If we need more minerals the best thing to do is flood the atmosphere with

CO_2 and melt off all the ice." That's a slow process, however. Faster ways to melt the ice will undoubtedly be found. Of course, any large-scale move would be dangerous. Aside from the climatic changes that would come about—so complex and unpredictable that one cannot realistically speculate about them—there is the matter of rising sea level which would occur as the ice melted.

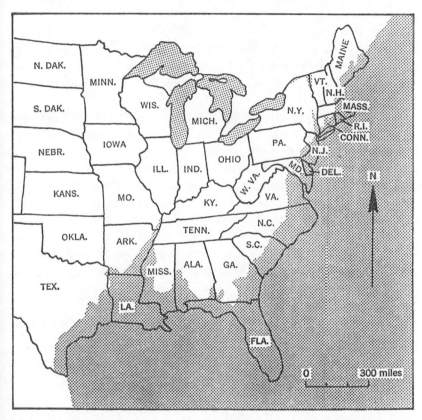

If all the world's ice were to be melted, the rise of sea level would move the present coastline of low-lying eastern United States inland, possibly to the position shown here. The gray area would be submerged. (U. S. Geological Survey.)

If the ice on Greenland alone melted it would raise sea level about 30 feet. That is enough to ruin waterfronts and make Venices of coastal cities and submerge much valuable rural land. If all the world's glacial ice melted, including that of Antarctica, sea level would rise between 140 and 250 feet, according to various estimates. Then much of the world's population would be displaced because some of the most densely populated places are deltas, coastal plains, and port cities. If the 250-foot mark were reached, the Statue of Liberty would be submerged except for her head and upraised arm. Florida, Louisiana, and Delaware would be virtually gone. A wide estuary would extend up the Hudson River valley to Albany and up the Mississippi Valley to southern Missouri. The Central Valley of California would be an inlet of the sea. There would be a gain of about 10 per cent of new, ice-free land, but there would be a loss of about 20 per cent of present land area which would be submerged.

We humans have already upset animal communities, vegetation, landforms, soils, lakes and rivers, and even things as vast as the atmosphere and oceans. Only glaciers seem to have escaped us. They are the least disturbed, most nearly natural of the major features on Earth. But they may be next. We see how cold lands are becoming valuable real estate; how mining, industry, tourism, agriculture, and other activities are increasing there; how formerly untouched zones of ice on land and sea are being closed in upon.

Strangely, or naturally, for various reasons more and more of Earth's inhabitants are moving into frontier areas that have a climate and environment similar to that which existed in our latitudes during cold stages of the Ice Age.

Maybe the Ice Age has been maligned. Maybe it was not so bad after all.

CHAPTER 25

THE FRIENDLY ICE AGE

O Winter! ruler of the inverted year, . . .

I love thee, all unlovely as thou seem'st,
And dreaded as thou art.

WILLIAM COWPER

Though winter is represented in the almanac as an old man, facing the wind and sleet, and drawing his cloak about him, we rather think of him as a merry woodchopper, and warm-blooded youth, as blithe as summer.

HENRY DAVID THOREAU

THE ICE AGE was hard on the land, for the glaciers gouged and scoured it and left it littered with gravelly waste. The Ice Age was hard on animal life, in that it saw the extinction of many species, including some of the most magnificent large beasts that ever lived. It was hard on plant life, in that vegetation patterns were continually disturbed as the ice waxed and waned. The Ice Age was hard on all those things, but as it turned out, it was a friend to humans. They too were pushed about, caused to migrate and readjust, and many types of Early People and near-people faded from the scene, even as many types of flora and fauna did; but with every ebb and flow of the ice sheets Early People grew in skills and intelligence until the last glaciation produced people like us. The human race thrived and progressed in spite of the Ice Age. Or did it do so *because* of it?

Reviewing the past, we see that millions of years ago, while glaciers were establishing themselves and Earth's geography was in an unusual state of flux, man-apes in the tropics were experimentally making things with their hands and starting to think abstract thoughts in a way animals never do. Then the great ice sheets plastered the North . . . and melted off . . . and came again—how many times we cannot be sure. By then the man-apes' far descendants —or ascendants—were in the middle latitudes with fire and superior Stone Age implements, acclimatized, and culturally prepared to live in the cold.

If the Ice Age cold had been detrimental to Early People, then those populations near the ice sheets should have been less advanced than those in Africa, where climate was always warm and they were not subject to the severely low temperatures felt in the North. One would think that Early People of the tropics and subtropics would have had more success in developing their skills because they did not have to spend nearly as much time securing food, clothing, and shelter. Yet as the Ice Age went on, the culture of Africa's Early People stood still and people in the North were showing greater inventiveness and were multiplying. The big cultural steps forward were being made not in the tropics but in the colder realms. Eurasians, toughened by climatic changes, living even into the subarctic, were more progressive than tropical populations, and in the last glacial stage a major population expansion took place in northern Eurasia and extended through subarctic North America into the New World.

Geographer Karl Butzer writes of the "transfer of man's cultural and biological focus from the tropics to the middle latitudes" in his book *Environment and Archeology:*

If tool-workmanship be an index of cultural progressiveness, and site tool density an index of population size, then Africa would probably qualify as the major center of population and cultural innovation of early and middle Pleistocene times. But during the last interglacial, tool-craftsmanship found a new focus in Europe. . . . And population density, insofar as can be inferred from the evidence, achieved a new high during the European late Pleistocene. By this time Africa may have become a cultural backwater. . . .

Ironically, the European center of accelerated cultural innovation during the last glacial was found in the then prevailing forest-tundra and cold loess steppes, rather than in the warmer, temperate woodlands of the Mediterranean region. Assuming that the . . . cave art marked the heart of the European culture area, one could suggest that the forest-tundra of Europe had displaced the African savannas as a center of innovation.

Modern-day proof that glaciers are not harmful things to live near is found in Iceland. That island just south of the Arctic Circle is 12 per cent glacier-covered, and the average life expectancy of its inhabitants is among the longest in the world!

To Neanderthals the cold and ice were not unfriendly. During the Würm glacial stage Neanderthals stayed on in Europe as the ice sheet lay over Scandinavia, and were adapted to the cold environment. Their brains developed to a size as large as (and sometimes larger than) ours. They were expert hunters, and they observed rites and practised burials in a way that indicates they thought of the supernatural and of an afterlife. But then the climate warmed in the mid-Würm, and one would expect that without the glacial "handicap" the Neanderthals would have progressed even further and faster. Instead, when it warmed they disappeared from the picture.

Upper Paleolithic people supplanted them and then the cold returned in late Würm, and the ice sheet spread into Germany, Poland, and Russia. These new people developed the finest quality and greatest variety of implements made before the use of metal, and clothing comparable to that of hunting people of the Far North in modern times. They too were consummate hunters, and their cave art equaled present-day art. Their culture had advanced during the coldest part of the last glaciation. But did it continue advancing when the ice sheet finally and completely withdrew? No, it degenerated as Europe warmed. The post-glacial population did not carry on their splendid art tradition, and it died.

So conditions in glacial times could not have been overly oppressive. Even the inhabited areas closest to the ice, on the tundra, may not have been as unpleasant as we picture them. As far as vegetation is concerned, they would have looked like the tundra of today along the Arctic Ocean, but they would have had a different feel, a different illumination. Those Ice Age tundras were at lower latitudes than those where Eskimos and Laplanders live today. They were not in the land of the midnight sun but at the latitude of Paris and Chicago, where the sun was much higher and stronger and the wintry days longer. Having brighter sunlight, the scene was more cheerful and probably more healthful; and animals were more numerous, for the carrying capacity of the land would have been greater at those latitudes.

Even a glacial stage had its summers, and although they were brief and cool they were a respite from the deep winter. In summer the tundra was a colorful riot of flowers.

Air masses draining off the ice caps and ice sheets were bitter cold, it is true, and winds strong with little to slow their speed, but the air must have warmed considerably in its descent. From the top of an ice mass thousands of feet high, descending air would have warmed adiabatically and would have had a considerably higher temperature at the base of the ice than it had when it left the top.

Warm, dry foehn (föhn) winds of the Alps surely were experienced at times. Foehn winds occur in most mountain areas under certain conditions when winds blow across mountain ranges into lowlands. These winds are warmed by heat released when condensation occurs as they go up over the mountains, and by adiabatic compression as they descend on the lee side. They bring relief from cold spells and "eat up" snow. Temperatures rise markedly with the coming of a foehn wind, and snow melts rapidly and is quickly evaporated due to the wind's dryness. Chinook winds on the east side of the Rocky Mountains are of the foehn type, and they would have been felt by Ice Age people too. When a chinook sets in after intense cold, the temperature may rise 20° to 40° in fifteen minutes. A foot of snow may disappear in a few hours. Such warm interludes must have helped make the usual cold more tolerable in some areas. Of course, not everyone lived in the colder regions near the ice. More southerly places had less harsh climates and more protected environments.

If the climate was rugged during glaciation—and it was—so were the people who stayed and lived in it. They were free to travel, in some ways freer than we. Many went south in winter or for life, but many did not. They were not as soft-muscled and short of breath as most of us half-sedentary, appliance-and-machine-reliant people of today are. They were at home in the out-of-doors as much as in their cave or shelter, and had to be in good physical condition at all times.

A cave or shelter under a ledge could have been quite cheery and comfortable, especially if it had a southern exposure. When the sun shone and wind did not blow directly in, it had a mild micro-climate. You know how on certain winter days—usually the coldest ones which are sunny with a north wind—you can open sunny south windows wide and suffer not a bit from the outside cold, and even enjoy the invigorating air. This pleasure they must have had often.

Caves, rock shelters, and tent-like huts of bones, branches, and hides seem inadequate and primitive to us, but they obviously were sufficient for Ice Age people. Some caves and rock shelters in Europe were occupied by a succession of people of different eras and cultures, for no one knows how long. In France today modern houses have been built under rock overhangs, houses whose fronts and sides are regular walls in lieu of rustic windbreaks but whose rear walls are the bedrock itself, the same kind of wall Neanderthals and other Ice Age people had in their "homes." Some southern Europeans find it fashionable to use a cave with a good view as a vacation residence which, when furnished, is at least as comfortable as a tent. And in the Loire Valley some man-made caves carved in soft bedrock are used as regular all-year homes. How close their occupants are still to life of the Ice Age!

Many Ice Age people lived fairly well even by our standards, and modern people are not completely divorced from the old ways. Although we wear the garb of civilization, something of the primitive remains in us. Along with our growing urbanization there is a compensatory clamoring to de-urbanize, to simplify life, to get away from cities, and to rough it—in essence, to live more as people did during the Ice Age, in the out-of-doors in a close-to-nature way. It is as though we were trying to regain some of what we have lost.

When hunting season starts, highways are crowded with cars leaving town, and hunters are glad when the ground is snow-covered, for it is then easier to track animals. Fishing goes on even in winter through holes cut in the ice, and fish caught from the cold waters have exceptionally firm flesh and fine flavor. Going "native," people enjoy living as Early People did—cooking over an open fire, picking berries, nuts, and wild herbs (food gathering), sleeping under the stars, hiking in the fresh air and wilderness with a few belongings slung over their shoulders. They find clothing style unimportant, and think nothing of doing without kitchen and bathroom plumbing.

Was the Ice Age really so bad? Is winter? Before you groan "yes," think what it is about winter that makes it objectionable. Are not most of its inconveniences the result of our artificial way of living?

We design our streets and our routines for summer weather, and then grumble when snow comes and we cannot get around as easily. The trouble with winter is not the weather exactly. It is the automobile problems; slippery sidewalks, because we are forced to walk where everyone else does; the jobs and schools we have to get to at different times; and going to and from work in the coldest parts of the day, often in the dark. If we lived under Ice Age conditions with no schedules, no transportation problems, no shopping, no store or office hours, no narrow skirts, high heels, and fancy hairdos, winter life would be far simpler and more comfortable in many ways.

If you had lived in an Ice Age winter thousands of years ago you probably would have slept more in the low-sun months and kept snug much of the time in your shelter next to your fire, listening to the crackling of fragrant burning wood. You would have kept frozen food and fuel nearby to use during especially bad weather, and stayed inside during snowstorms and the coldest cold waves and when the ground was slick; and when conditions improved you would have hunted. When you went out you would have been voluptuously clad in luxurious furs that today's richest ladies would envy. You would have walked around drifts instead of shoveling them.

These people did not fight snow, ice, and cold as we constantly do. They adjusted to them.

Many people sincerely like winter—the invigorating briskness, the glitter of icicles, the smooth contours of the snow. Coldness gives good tone to the body, and no light is more flattering to the face than the well-scattered light reflected from pure-white snow. The cleanness of snow and freezing cold together rids the environment for a season of insects, rank vegetation, and dust. Landscape clutter is hidden and all outdoors is given a brand-new look.

Now that our affluence has given us more leisure time, we are finding that winter really can be more enjoyable than we have let it be. In the past when city dwellers acquired a second home it was usually a summer home, but now it is often a winter home as well. More vacations are taken in the winter now because of the growing popularity of skiing, snowmobiling, and other winter sports. The number of winter visitors to national parks has increased considerably.

Not all birds fly south in winter, and not all people do either, or would want to. Many go north.

No, winter is not as hostile or unfriendly as it is reputed to be, and neither was the Ice Age.

As we have seen, much of the Ice Age was not glacial at all. During interglacial periods climate became as warm as and warmer than now. Even during the glacial stages, not all the world was cold. Much of it hardly felt a drop in temperature. And some regions that are desert now were more humid and green during parts of the Ice Age, some even forested.

It is true, while the ice sheets existed the land they covered was useless, but there was compensation, for as they enlarged sea level fell and continental shelves and other shallowly submerged parts of the ocean floor emerged as inhabitable land. Land bridges appeared, providing avenues of travel between continents and islands.

Loren Eiseley writes, "We are the offspring of the sleeping ice." And look at some of the legacies it left us.

It left us some of the most beautiful and inspiring of the world's scenery. In mountains the glaciers gave us artistic horn peaks, cirques, jewel lakes, comb-like ridges, and tall waterfalls that hang down steep cliffs like tinsel on Christmas tree boughs. The wide, flat-floored, U-shaped valleys they carved and then carpeted with drift give mountainous areas broad strips of level land for villages, farms, and transportation routes. Fiords give ships protected harbors and deep-water entrances into many mountainous areas.

One cannot begin to mention all the well-known lakes that owe their existence to the glaciers' scouring and deposition. Most of the really large lakes of the Northern Hemisphere—Lake Winnipeg and Great Bear and Great Slave lakes in Canada, Lake Ladoga in the USSR near Finland, and of course the Great Lakes, which serve as a valuable inland waterway to the heart of North America and which hold one fifth of the earth's fresh surface water. The long, narrow Finger Lakes of New York State, pointing the way the ice went. The numerous picturesque lakes scattered through glaciated mountains, lying like crystal beads of a broken necklace connected loosely by the strings of streams. Beautiful Lakes Como, Garda, Maggiore, and others in northern Italy, which formed behind moraines of the Alps glaciers. And those of England's Lake District, which was an inspiration to Wordsworth, Coleridge, De Quincey, Ruskin, and many other literary figures. The uncountable thousands of smaller lakes in the northern parts of Minnesota, Wisconsin, and Michigan, and those in New

England. Parts of Canada and Finland have more water surface than land, and one can travel almost endlessly there by canoe with easy portages.

Glaciation left ponds and marshes too which are important water reservoirs and wildlife havens. Marshes used to be thought of as useless, unsightly things, blemishes that should be removed from the landscape. Now, however, we appreciate their function in controlling water levels and harboring wildlife, and—yes—their beauty. In countless communities citizens are now campaigning and fighting to have their local marshes preserved.

Gravelly glacial drift is a porous sponge capable of holding large quantities of water. Wells in deep drift usually have a copious, reliable water supply.

Glacial gravel and sand are of immense economic value, being used for concrete, earth fill, and road materials. Highways in areas of glacial deposition are usually more cheaply built and better maintained than those in most nonglaciated regions because of the widespread availability of gravel which builds sturdy, stable, well-drained roadbeds. Pure, fine sand that was sorted by meltwater is used by industry in molds for metal castings and in making glass.

Old, empty glacial spillways cut by rivers of meltwater that no longer exist have been good routes along which to dig canals.

Over large areas glacial deposition improved the land for agriculture, smoothing its surface and renewing its fertility. The ice sheets eroded away a thick mantle of old, residual rock material and soil which had been leached of minerals that plants need to grow well. In its place it left a new layer of drift that contains a mixture of fresh mineral matter. The gravelly drift of till plains is made up of pieces of rock over which the glacier passed—many different kinds ranging in size from large boulders to rock flour, with a variety of nutrients. Fertility is continuously restored because year after year the larger rock particles keep breaking down and weathering and releasing nutrients to the soil.

Louis Agassiz, one of the greats among the early glacial scientists, referred to the ice as "God's great plough" in his *Geological Sketches*, in which he wrote in 1876:

What was the use of this great engine set at work ages ago to grind, furrow, and knead over, as it were, the surface of the earth? We have our answer in the fertile soil which spreads over the temperate regions of the globe. . . . The hard surface of the rocks was ground to powder, the elements of the soil were mingled in fair proportions, granite was carried into the lime regions, lime was mingled with the more . . . unproductive granite districts, and a soil was prepared fit for the agricultural uses of man.

Much of the world's best farms are on land the ice sheets deposited—level or rolling till plains, smooth outwash plains, flat floors of former glacial lakes and marshes, and in areas of loess soil blown by glacial winds.

Whereas the glaciers smoothed many hilly areas by filling deep valleys with drift, they added variety to other landscapes by building interesting irregularities—drumlins, kames, eskers, crevasse fills, and moraines. Moraines are choice recreational sites with ponds and lakes and hummocky terrain. Most moraines have not been farmed but were left in trees because of their rough and often bouldery surface. Now they are in demand as sites for homes, estates, small ski hills, parks, and other recreational facilities.

Erratic boulders may exasperate farmers but they are often prized as decorations and are utilized by landscapers. They have been used as construction material too for thousands of years. Numerous "fieldstone" buildings in glaciated northern United States, Canada, and Europe (including some of Europe's oldest castles) have been built of hard glacial boulders. As farmers cleared their fields they usually used the boulders to build their homes, barn foundations, and fences.

Is it just coincidence that those parts of northwestern Europe and east-central North America across which the fronts of the ice sheets fluctuated so many times, where they held sway so long and deposited their richest gifts—that those regions came to be world leaders in agricultural and industrial production, commerce, the arts, and world affairs? Other factors are also responsible, of course, but the effects of glaciation cannot be overlooked.

The Ice Age is usually thought of as a dismal twilight when the earth sank into a deep freeze and suffocated under a deadening hood of ice—a kind of Dark Age shrouded in gloomy stagnation. In certain places, especially northern North America and northern Europe, it was exactly that—much, but not all, of the time.

Just as with every calamity there comes some good, with every setback some reward, so did the earth benefit from its temporary affliction. While it is true that the lethal ice wrought tremendous havoc, still one should not think of it entirely as a scourge any more than one thinks that of a winter, no matter how severe. Rather, one should think how adversity stimulates resourcefulness and evokes unrealized strengths and miraculous responses, how even in winter the cold, moist ground freshens and greens even under the snow, and how all living things stir impatiently, waiting to contribute their reviving energies to the exuberance of a new spring.

During the Ice Age the glaciated world was not dead. It was in many ways just coming to life.

CHAPTER 26

WHILE THERE IS TIME

We never seem to know what anything means till we have lost it . . . till in place of the bright, visible being, comes the awful and desolate shadow where nothing is—where we stretch out our hands in vain, and strain our eyes upon dark and dismal vacuity.

ORVILLE DEWEY

IT IS human nature to take life's good things and good times for granted until they are gone, and to be unaware ofttimes that one is experiencing a memorable event until it has passed. How many people who lived through critically important episodes of history were insensitive to their significance? How many precious things in our personal lives are regarded lightly until they are no more—a close friend, congenial neighbors, our parents' voices, our youth, our middle age, good health at any age.

We live too late to see the Jurassic jungles and their dinosaurs, too early to visit the planet Venus, but we are on time to experience part of the Ice Age, which may be its grand finale. We can still see it face to face while some of its royal splendor remains. We can still touch the hem of its trailing toga before it exits, read its torn scroll, and feel firsthand the sensation of its presence.

What remains of the Great Ice Age for us to see is, of course, two powerful, Pleistocene-type ice sheets and a multitude of glaciers of sundry sizes and types. But that is not all. There is also the ice's erosional and depositional work upon the terrain. There are the resources glaciation left. There are animals that survived the glacial epoch. There are whole physical environments that have slowly been changing as the ice contracted or disappeared.

Knowing that fluctuation has been the habit of the ice, we realize it could grow back in full force in the future. Its cold, sticky fingers still hold our globe, pressing on various spots as though it will not let go and might grab it tighter again.

We know that a number of times in the past the ice retired as it is doing now and waited for its cue to come forth again, as a star in her dressing room waits to come on stage for another act. And quite a star the ice has been. What has surpassed her performance and pageantry? The intermittent pyrotechnics of a volcano? The pirouette of a whirling hurricane? The lazy lifting of a mountain? The silent sinking of a sea? All such grand phenomena are outdazzled by the spectacle of the moving ice. With encore after encore she reappeared and receded rhythmically, spreading out her shiny scalloped skirt and then drawing it back. If she makes another appearance it will be far in the future, beyond our lifetime and vision. As far as we are concerned, her famous acts are things of the past.

From our position in time the Ice Age lies behind us, and we see the remains of that extraordinary, exciting period being rapidly erased, physically and esthetically. Many glacial features have been totally removed; some are covered over or remodeled beyond recognition; others, and many glaciers themselves, are so disturbed that their naturalness is gone.

The speed at which the human race is ravaging the planet is electronic. Man's compulsion to conquer wilderness and carry civilization into every nook of the world continues, and his ability to do so keeps increasing.

Glaciers—and only some of them—are among the few chaste, untainted natural features on Earth. Soon no glaciers may be that way, for the spread of exploration, settlement, resource development, pollution, and mechanization goes on energetically. We have seen how our iceward migration is progressing; how towns, cities, military installations, and mining and industrial enterprises are moving into cold regions; how sportsmen and tourists are zeroing in on glaciers; how would-be modifiers of climate and environments are planning how to get rid of ice; and how glaciers are melted when more water is needed.

"But," someone says, "there is so *much* ice on this planet! Why worry if some is melted or loses its natural beauty?" It is felt that if some glaciers do suffer from misuse and careless management of the environment (which have struck practically every other natural feature you can think of) there will always be more clean, glisteningly beautiful glaciers somewhere—in the polar areas if nowhere else.

However, those "frigid zones" are not as vast as we tend to think. Compare the area of tropical, temperate, and polar zones. Of the whole surface of the globe—197,000,000 square miles—almost 80 million square miles lie within the tropics, the wide belt around the world's fat waist. The two temperate zones together cover over 100 million square miles. And though there are two polar zones too, both of those together comprise only 16 million square miles—just one fifth the area of the tropics! And the north polar zone is, of course, largely ocean. Development of cold regions is going on so fast that they are likely to be opened before many tropical areas are. Eskimos already are far more educated, mechanized, and internationalized than Indian tribes of the Amazon rainforest and inhabitants of many other tropical regions. Glaciers are not safe from exploitation and defilement no matter where they are, and though they seem endless in number, they are not.

We have pointed out how many vital earth resources will soon become exceedingly scarce or exhausted as population growth continues unchecked; how underdeveloped countries are striving for greater food production and industrialization; how fresh water is already so urgently needed in some areas that it is transported long distances through redirected rivers and artificial watercourses; how new mineral deposits are eagerly sought as the world's living standards rise and backward countries develop. We have seen how melting of glaciers would provide more fresh water (temporarily) for agriculture, power, industry, and domestic use; would expose more minable resources which are now hidden; would free more harbors; and in some localities would lessen flood dangers and "improve" climate. The ice-must-go and warm-the-environment advocates will speak out more forcefully as time goes by.

We have seen too that, as the population proliferates, the area per person of wilderness and natural recreational land becomes smaller. You personally know you have less elbow room, less living space, less of the peaceful, clean out-of-doors to yourself than you had even a few years ago. We feel the physical and psychological effects of overcrowding, and this press is moving into glacier regions.

It stands to reason that when our lakes and rivers have become polluted, when even our oceans are threatened, when much of our good soil is ruined or blown away, when our virgin forests are slashed down—even our last redwoods—when the atmosphere is made sickening, when much of our wildlife is being driven or hunted to extinction—then glaciers in their heretofore protected sanctuaries will not long remain inviolate. It is not the way of our world that they could.

Fall-out from modern civilization's air pollution and stirred-up dust is darkening glaciers around the world—ever so slightly in some cases, noticeably in others. Such "dusting," if continuing, could accelerate the melting of some or all glaciers, even if general temperatures did not rise, and conceivably even if they dropped.

This period of history—this impatient, reckless scientific age—may be the last chance to see pristine glaciers. As for other legacies of the Ice Age, many of those have already changed or disappeared.

Even without Man's earth-moving machines, natural processes of gradation are always at work filling in low spots and wearing down high ones, and so landform features are eventually reshaped.

Glacier-made lake basins, like all lake basins, disappear in time. An out-flowing stream wears down the lake's rim allowing the water to drain away, or the lake fills gradually with sediment that washes in and with the remains of vegetation that grows and dies in it. Lakes fill more quickly when domestic waste material is discharged into them and when vegetation growth increases as a result; and when plowing, deforestation, and other land-disturbing activities speed soil erosion and increase the amount of sediment washed into the basins. As vegetation grows in around the edges of lakes they become smaller and shallower, turning into marshes. Marshes eventually fill in completely and become just flat land. Many marshes in glaciated regions are former lakes that are now in a partially filled state. Artificial draining is hastening their disappearance.

Natural erosion, farmers' plows, road builders, and commercial sand-and-gravel companies are wearing down, altering, or completely eradicating many irreplaceable features of glacial deposition—kames, eskers, crevasse fills, drumlins, terraces, kettle holes (where ice blocks once stood), the gently undulating swell-and-swale topography of till plains. In glacial deposits, especially those built by glacial streams, gravel companies find their materials already well sorted, and they haul sand and gravel away by the truckloads and trainloads. Many kames, eskers, and other glacial features have been entirely removed in this manner. Pits are often dug in drumlins or moraines for as much gravel as is needed and then the unsightly holes are left, frequently to be used as community dumps. Streamlined drumlin hills, which usually occur in clusters, have roads cut through them, and the gravel of which they are built is used to make roadbeds and fill low spots. It is not "wrong" to do these things. Resources must be used; roads must be built. But it is "right" that we preserve some of the vanishing glacial landscape. It is already too late to save some of the best examples.

Boulders dropped by the melting ice have been disappearing in populated areas. As you drive through much of the glaciated country that has been settled and farmed for some time—the southern Great Lakes region or northern Great Plains, for example—you have to look hard to see glacial boulders. Occasionally you see them in ditches into which they have been pushed. Farmers have hauled them from fields to the fence lines and hedgerows, where they lie overgrown with brush and weeds. They are easier to see there in winter when vegetation does not hide them as much. Beautiful, erratic boulders have been rolled into excavated holes and covered up by highway crews, builders, and homeowners. They are all but invisible in urban areas. The rare, exceptionally large ones, which would be dramatic reminders of the ice's power if placed where the public could see them, are often blasted apart in order to dispose of them. And all boulders, wherever they are, are slowly weathering down to smaller size. Some glacial erratics were set on rock outcrops with unbelievable gentleness by stationary, melting ice in a most unnatural, top-heavy position. They sit perched, like art objects on pedestals, unstably balanced in a way that tempts one to give them a shove and make them topple down. The shove has been given to countless numbers of these perched boulders, and will be given to many more. And then they lie humbly on the ground among the others. Many are so huge that no machine could possibly set them back the way they were.

Weathering and erosion are wearing down and smoothing away all the features left by the ice, not only its deposits but also its sculpturing in hard rock. Just after the ice left, the scratched lines and carvings it made were still fresh. But like the engraved stone of ancient buildings and obelisks, subject to rain, frost, and wind blasting, they have been losing their sharpness. The fine, delicate work was first to go— thin striations and groovings; the high-gloss polish; and chattermarks, little crescent-shaped notches chipped by rocks being jerked along at the base of the ice. Larger features like cirques, horns, and troughs will remain longer, many thousands of years longer, but their lines will soften, their contours will become indistinct, and in time they too will be worn away.

Other legacies of the glaciers are being lost. Idly wandering rivers and streams, whose courses were picturesquely disarranged by the ice and drift, are straightening out into ordinary, more direct channels. Sandbars in valley trains are washing down to the sea. Waterfalls and rapids are wearing down. Fossil water stored from the Pleistocene in underground rock of arid regions is being tapped and will not be replenished, for rain is too scanty there now. Bones of Ice Age animals, where found in quantity, have been ground up for fertilizer. Mammoth tusks have been collected by ivory traders and sold to be carved into ornamental and utilitarian objects. About 25,000 mammoths have been unearthed in Siberia. Before World War I about 32 tons of mammoth-tusk ivory were exported from northeastern Siberia annually. Each year there are fewer tusks to be found. The Ice Age artwork in caves chips off and is rubbed off, and the green fungus creeps over it. Many lakes that turned salty when the Ice Age ended—like Great Salt Lake in Utah—continue shrinking or have become just salt flats. Surviving species of big game, magnificent large birds, and marine animals are still hunted down or are losing their habitats, and many live dangerously near extinction. Our larger, more spectacular animals are being killed off first. Many of our animals are living fossils of the Pleistocene; having gone through perilous environmental adjustments, they may vanish before our eyes: whales, caribou, muskoxen, polar bears and other big bears, buffaloes, the great apes, elephants and rhinoceroses which once roamed round the world, and many other disappearing animals seen in zoos.

Glaciers themselves are always traveling downward to their death. Fortunately there are still all kinds of them for us to see—some brilliant white; some veiled in gray mist, barely visible through clouds and fog; some frighteningly powerful; some pitifully eroded or fragmented. We can still walk on plains and hills they fashioned and view their massive mountain sculpturing. We can hold in our hands rocks they smoothed and scratched, and sail on lakes and canoe on rivers whose settings they created.

The Ice Age fortune has been bequeathed to us for our lifetime, and how we handle it determines how much of it will be left for future generations. It is as though the funeral of a wealthy relative has taken place and the heirs will be bickering over the handling of the estate. Some family members cherish handed-down possessions and out of love try to preserve as many as they can; and they realize that what is scattered and squandered is gone forever. Others would liquidate the estate without any qualms. They would dispose of the old homestead, auction off the antiques, sell the family jewels to the highest bidder, carelessly use irreplaceable heirlooms, have the pets put away, and feel no remorse. Our glacial inheritance could be frittered away with as little thought and feeling.

We cannot keep it all intact, nor would we want to. Just as money has to be invested and put to use, so resources left from the Ice Age are to be used, but wisely, for mankind's needs and enjoyment. But we should also cherish some of our inheritance and preserve it for those who follow us.

If enough devoted members of the human family speak up in time, part of this vanishing legacy can be kept for coming generations. There is time for environment-conscious citizens of the world to see that Ice Age flora and fauna are protected, that at least some glacier-made features and landscapes and some entire glaciers are allowed to remain in their natural condition.

Sometimes in this book we may have stressed certain points rather heavily to support our theme, to stimulate appreciation of glaciers and the Ice Age. If so, this means of emphasis should be condoned. Orators wax eloquent to get their message across; musicians are expected to play loudly at times to stir the listener; artists may use intense colors and work in bold relief to achieve their effects; novelists and dramatists portray their scenes and characters "bigger than life." Powerful language and powerful thoughts should be used to tell the story we have told, and maybe we have not written boldly enough.

Think back in time. And learn again to like snow and coolness and even the bitter cold. When a cold front comes do not cringe and shiver as you have been trained to do. In summer feel how your body drinks in its refreshing relief after the sweltering heat. In winter too breathe deeply of the invigorating air and realize that this is one of the stimuli that made our species strong and successful. Hold up your head on a winter's day, stride strongly into the wind, and feel the charge your mind and body receives. It is what you were missing and needing all summer.

And when the snow falls do not listen to the complaints of others, but see it for what it is, as you did when you were a child—one of Earth's most resplendent decorations, glittering with sparkles, formed in intricate shapes. Think of it as composed of the life-giving moisture we need. See how it adds beauty to everything it lies on, how its graceful curves and shadows take on pastel colors at dawn and twilight, and how it brightens the night.

Pablo Picasso designed the 162-ton modernistic steel sculpture, five stories high, that was erected in the plaza of Chicago's civic center in 1967. He was eighty-six years old, living in France, and unable to see the creation when it was finished. Speaking with his deep appreciation of beauty and composition, this great artist said, "Be sure to send pictures when the snow falls."

How much more beauty we could find in this world if we saw snow through the eyes of the artist instead of the snow plower.

And when you think of glaciers, do not think of them as objects to do battle with. Think of them as friends, friends we grew up with.

The Russian geographers Avsyuk and Kotlyakov wrote this about glaciers: "It has been said that glaciers are a negative phenomenon of nature, a global 'malady.' . . . We do not share this view. Glaciers are a source of health to our planet since they help to shape climatic contrasts and thus the rich diversity of our global environment."

If glaciers were not on this planet, adorning and enlivening it and regulating its climate, this would be a much different and less congenial world-home for us.

Ice Age Lost has been a survey of the Ice Age from one person's point of view, using past, present, and future perspectives. It is a collection of facts as they are currently known. It is not the full story, for that cannot be told by any one writer and certainly not in one book. This is a general picture as seen at this time, but it will change soon, and probably markedly, as new facts are found by researchers in many fields and as new interpretations are made from the growing pool of knowledge.

The Ice Age has seemed far distant from us, but we are learning it is nearer than we thought. More people by the thousands are living in true Ice Age environments, in Antarctica and Greenland, than were there a few years ago. Scientists are uncovering and piecing together masses of data about Ice Age geography and are making new discoveries about the Early People who lived then. Conservationists have helped preserve for us much of the wildlife, vegetation, and wilderness left from the Ice Age. Improved transportation is making it possible for us to visit glacial regions in almost any part of the world, and to experience Ice Age conditions in varying degrees.

The past is releasing its secrets little by little. And though many of them will never be known, we have reason to hope that our Ice Age is not, and will not become, an Ice Age largely lost, but rather an Ice Age that is rediscovered and well remembered.

SELECTED REFERENCES AND SOURCES

THE ICE AGE IN GENERAL

Butzer, Karl W. *Environment and Archeology: An Ecological Approach to Prehistory.* Chicago: Aldine-Atherton, 1971.

Charlesworth, J. K. *The Quaternary Era.* 2 vols. London: Edward Arnold, 1957.

Flint, Richard Foster. *Glacial and Quaternary Geology.* New York: John Wiley & Sons, 1971.

Kurtén, Björn. *The Ice Age.* New York: G. P. Putnam's Sons, 1972.

Schultz, Gwen. *Glaciers and the Ice Age.* New York: Holt, Rinehart & Winston, 1963.

West, R. G. *Pleistocene Geology and Biology, with especial reference to the British Isles.* New York: John Wiley & Sons, 1968.

Wright, H. E., Jr., ed. *Quaternary Geology and Climate.* Washington, D.C.: National Academy of Sciences, 1969.

——, and David G. Frey, eds. *The Quaternary of the United States.* Princeton, N.J.: Princeton University Press, 1965.

EARLY PEOPLE AND THEIR ENVIRONMENT

Augusta, Josef, and Zdeněk Burian. *Prehistoric Man.* Trans. Margaret Schierl. London: Paul Hamlyn, 1960.

Chard, Chester C. *Man in Prehistory.* New York: McGraw-Hill Book Company, 1969.

Constable, George. *The Neanderthals.* New York: Time-Life Books, 1973.

Cornwall, Ian W. *The World of Ancient Man.* New York: The John Day Company, 1964.

——. *Ice Ages: Their Nature and Effects.* New York: Humanities Press, 1970.

Edey, Maitland A. *The Missing Link.* New York: Time-Life Books, 1972.

Hopkins, David M. *The Bering Land Bridge.* Stanford, Calif.: Stanford University Press, 1967.

Howell, F. Clark. *Early Man.* New York: Time-Life Books, 1971.
Howells, William. *Mankind in the Making.* Garden City, N.Y.: Doubleday & Company, 1967.
Laming, Annette. *Lascaux.* Trans. Eleanore Frances Armstrong. Baltimore: Penguin Books, 1959.
Leakey, Richard E. "In Search of Man's Past at Lake Rudolf," *National Geographic,* May 1970.
———. "Skull 1470—New Clue to Earliest Man?," *National Geographic,* June 1973.
Montagu, Ashley. *Man: His First Two Million Years.* New York: Columbia University Press, 1969; New York: Delta, 1970.
Pecora, William T., and Meyer Rubin. "Absolute Dating and the History of Man," *Time and Stratigraphy in the Evolution of Man.* Pub. 1469. National Research Council, Division of Earth Sciences, Washington, D.C., 1967.
Pfeiffer, John E. *The Emergence of Man.* New York: Harper & Row, 1972.
Robinson, John T. *Early Hominid Posture and Locomotion.* Chicago: University of Chicago Press, 1972.
White, Edmund, and Dale Brown. *The First Men.* New York: Time-Life Books, 1973.

ICE AGE ANIMALS

Augusta, Josef, and Zdeněk Burian. *A Book of Mammoths.* Trans. Margaret Schierl. London: Paul Hamlyn, 1962.
———. *Prehistoric Animals.* Trans. Greta Hort. London: Paul Hamlyn, 1960.
Kurtén, Björn. *Pleistocene Mammals of Europe.* Chicago: Aldine Publishing Co., 1968.
———. *The Ice Age.* New York: G. P. Putnam's Sons, 1972.

GLACIERS AND ICE

Avsyuk, Grigori, and Vladimir Kotlyakov. "Glaciers on the Move," *UNESCO Courier,* June 1969.
Dyson, James L., chmn. *Glaciers of the American Rocky Mountains.* New York: American Geographical Society, 1952.
———. *The World of Ice.* New York: Alfred A. Knopf, 1962.
A Functional Glossary of Ice Terminology. Pub. 609. Washington, D.C.: U. S. Navy Hydrographic Office, 1952.
Harrison, Arthur E. *Exploring Glaciers with a Camera.* San Francisco: Sierra Club, 1960.
Mellor, Malcolm. *Snow and Ice of the Earth's Surface.* Hanover, N.H.: Cold Regions Research and Engineering Laboratory, 1964.

Mercer, John H. *Southern Hemisphere Glacier Atlas.* New York: American Geographical Society, 1967.

Post, Austin. *Effects of the March 1964 Alaska Earthquake in Glaciers.* Geological Survey Professional Paper 544-D. Washington, D.C.: U. S. Government Printing Office, 1967.

——, and Edward R. La Chapelle. *Glacier Ice.* Seattle: University of Washington Press, 1971.

Ragle, R. H., J. E. Sater, and W. O. Field. *Effects of the 1964 Alaskan Earthquake on Glaciers and Related Features.* Research Paper 32. Montreal: Arctic Institute of North America, 1965.

SURGING (GALLOPING) GLACIERS

Desio, Ardito. "An Exceptional Glacier Advance in the Karakoram-Ladakh Region," *Journal of Glaciology,* Oct. 1954.

Mason, Kenneth. "The Study of Threatening Glaciers," *The Geographical Journal,* Vol. LXXXV, Jan. to June 1935.

Mercer, John H. "Glacier Variations in the Karakoram," *Glaciological Notes,* American Geographical Society, April 1963.

Post, Austin S. "Exceptional Advances of Muldrow, Black Rapids and Susitna Glaciers," *Journal of Geophysical Research,* Nov. 1960.

"Spectacular Surges of Glaciers Reported," U. S. Geological Survey press release, Oct. 30, 1966.

Zabirov, R. D. "About the State of Some of the Tien Shan Glaciers during the Period of the International Geophysical Year," *Proceedings of the International Union of Geodesy and Geophysics, Snow and Ice Commission,* Helsinki, 1961.

MANIPULATION OF GLACIERS

"Artificial Glaciers," *Ice,* Apr. 1964 and Dec. 1964.

Avsyuk, G. A. "Artificial Intensification of the Melting of Mountain Glaciers to Increase the Stream Flow in Central Asia," *Soviet Geography: Review and Translation,* American Geographical Society, Feb. 1963.

Hahn, Fred D. *Interim Report.* Power Bulletin No. 2. Department of Conservation, Glaciological Research Program, State of Washington, 1961.

Price, Truman P. *Weather Modification: A Power Potential.* Bulletin No. 1. Department of Conservation, Division of Power Resources, State of Washington, 1959.

Wiens, Herold J. "Regional and Seasonal Water Supply in the Tarim Basin and Its Relation to Cultivated Land Potentials," *Annals of the Association of American Geographers,* June 1967.

THE CLIMATIC RECORD

Huntington, Ellsworth. *Civilization and Climate.* New Haven, Conn.: Yale University Press, 1924.
Ladurie, Emmanuel Le Roy. *Times of Feast, Times of Famine: A History of Climate Since the Year 1000.* Garden City, N.Y.: Doubleday & Company, 1971.
Lamb, Hubert H. *The Changing Climate.* London: Methuen & Co., 1966.
———. *Climate: Present, Past and Future.* 2 vols. London: Methuen & Co., 1972.
Ludlum, David M. *Early American Winters, 1604–1820.* Boston: American Meteorological Society, 1966.
Mitchell, J. Murray, ed. *Causes of Climatic Change.* Boston: American Meteorological Society, 1968.
Silverberg, Robert. *The Challenge of Climate: Man and His Environment.* New York: Meredith Press, 1969.
"Snow Survey of the British Isles," *Journal of Glaciology,* Jan. 1947.
Ward, William Hallam, ed. *Variations of the Regime of Existing Glaciers.* Symposium of Obergurgl. Pub. 58. International Association of Scientific Hydrology, 1962.

ARTIFICIAL MODIFICATION OF WEATHER AND CLIMATE

Battan, Louis J. *Harvesting the Clouds.* Garden City, N.Y.: Doubleday & Company, 1969.
Chamberlain, A. R., chmn. *Weather and Climate Modification.* Washington, D.C.: National Science Foundation, 1966.
Greenfield, S. M. *Weather Modification Research: A Desire and an Approach.* Santa Monica, Calif.: Rand Corp., 1969.
Halacy, D. S., Jr. *The Weather Changers.* New York: Harper & Row, 1968.
MacDonald, Gordon, J. F., chmn. *Weather and Climate Modification: Problems and Prospects.* 2 vols. Pub. 1350. Washington, D.C.: National Academy of Sciences, National Research Council, 1966.
Malone, Thomas, chmn. *Weather and Climate Modification.* Washington, D.C.: National Academy of Sciences, 1973.

ATTRACTION AND DEVELOPMENT OF COLD REGIONS

Bohn, Dave. *Glacier Bay: The Land and the Silence.* San Francisco, Sierra Club, 1967.
Debenham, Frank. *Antarctica: The Story of a Continent.* New York: The Macmillan Company, 1961.

Dort, Wakefield, Jr. "Camping in Antarctica," *Sierra Club Bulletin*, July 1968.
Hornbein, Thomas F. *Everest: The West Ridge*. San Francisco: Sierra Club, 1968.
Ley, Willy. *The Poles*. New York: Time-Life Books, 1969.
Matthews, Samuel W., and William R. Curtsinger. "Antarctica's Nearer Side," *National Geographic*, Nov. 1971.
Potter, Neal. *Natural Resource Potentials of the Antarctic*. New York: American Geographical Society, 1969.
Quam, Louis O., ed. *Research in the Antarctic*. Washington, D.C.: American Association for the Advancement of Science, 1973.
Schultz, Gwen M. "Was the Ice Age Really So Bad?," *Frontiers*, Academy of Natural Sciences of Philadelphia, June 1964.
Stefansson, Vilhjalmur. *The Friendly Arctic*. New York: The Macmillan Company, 1921.
——. *The Northward Course of Empire*. New York: The Macmillan Company, 1924.
Victor, Paul-Émile. *Man and the Conquest of the Poles*. Trans. Scott Sullivan. New York: Simon & Schuster, 1963.
Washburn, Bradford. *A Tourist Guide to Mount McKinley*. Anchorage, Alas.: Northwest Publishing Company, 1971.

HISTORICAL READINGS

Agassiz, L[ouis]. *Geological Sketches*. Boston: James R. Osgood and Co., 1876.
Byrd, Richard E. *Alone*. New York: G. P. Putnam's Sons, 1938.
Muir, John. *Travels in Alaska*. Boston: Houghton Mifflin Company, 1915.
——. *Stickeen: The Story of a Dog*. Boston: Houghton Mifflin Company, 1937.
Nansen, Fridtjof. *The First Crossing of Greenland*. Trans. H. M. Gepp. London: Longmans, Green & Co., 1892.
——. *Farthest North*. New York: Harper and Brothers, 1898.
Twain, Mark. *A Tramp Abroad*. London: Chatto & Windus, 1880.
Tyndall, John. *The Forms of Water in Clouds and Rivers, Ice and Glaciers*. New York: D. Appleton and Co., 1896.
——. *Glaciers of the Alps and Mountaineering in 1861*. London: J. M. Dent & Co., 1906.

OTHER REFERENCES

Black, Robert F. "Geomorphology of Devils Lake Area, Wisconsin," *Transactions of the Wisconsin Academy of Sciences, Arts and Letters*, vol. LVI, 1968.
Dobzhansky, Theodosius. "The Present Evolution of Man," *Scientific American*, Sept. 1960.

Harris, Chauncy D., ed. *Soviet Geography: Accomplishments and Tasks.* Trans. Lawrence Ecker. New York: American Geographical Society, 1962.

Park, Charles R., Jr., with Margaret C. Freeman. *Affluence in Jeopardy.* San Francisco: Freeman, Cooper, 1968.

Perutz, M. F. "A Description of the Iceberg Aircraft Carrier," *Journal of Glaciology,* March 1948.

SELECTED MAPS

Field, William O., and associates. *Atlas of Mountain Glaciers in the Northern Hemisphere.* Various scales. American Geographical Society, 1958.

Glacial Map of Canada. Scale: 1:5,000,000. Department of Energy, Mines and Resources, Geological Survey of Canada, Ottawa, 1968.

Glacial Map of the United States East of the Rocky Mountains. Scale: 1:1,750,000. Geological Society of America, 1959.

Glaciers of the Southern Hemisphere. Various scales. Glacier Studies, American Geographical Society, 1967.

Nine Glacier Maps, Northwestern North America. Scale: 1:10,000. Special Pub. 34. American Geographical Society, 1960.

Retreat of Wisconsin and Recent Ice in North America: Speculative Ice-Marginal Positions during Recession of Last Ice-Sheet Complex. Map 1257A. Scale: 1:5,000,000. Department of Energy, Mines and Resources, Geological Survey of Canada, Ottawa, 1969.

INDEX

Abrekkebreen Glacier, 170
Africa
  considered likely birthplace of
    Man, 20
  East, large animals in, during time
    of *Homo erectus*, 94
  glaciers in, 5
  man-ape fossils in, 84–85
  North, rain in, during glacial
    times, 91
Aftonian Interglacial Stage, 31
Agassiz, Louis, 310
Agriculture
  birth of, 279
  irrigation in, 143
  land needs of, 238
Air masses
  cold, grew with ice sheets, 55, 57,
    58
  collision of, storms caused by, 57
  from glaciers, induced precipita-
    tion, 57
  source regions of, 55
  tropical and polar, in conflict, 3
  warm, challenged advancement of
    ice, 56
Air pollution, 209, 316
  as climate modifier, 152, 216
  as smog over large cities, 226
Alaska
  glaciers of, 171–72, 262
  ice caps in, early, 52
  resources of, 287
  vegetation of, before Ice Age, 14

Albedo (of snow), 59
Aletsch Glacier, 130
Alps
  formation of, 16
  glaciers of
    decrease in, 143
    localized, 38
  ice cap over, 111
  ice-stage patterns in, 30
  local climates in, 142
  as water-distribution hub, 142
Altamira cave (Spain), 113, 271
Amu Darya River, 138
Andes Mountains
  building of, 16
  early ice caps in, 52
Animals
  in America during Ice Age, 119
  array of, prior to Ice Age, 14
  carnivores, preyed upon grazing
    herds, 18
  decreased in size, 120
  domestication of, 115
  in Europe, during interglacials, 98
  evolved ability to withstand cold,
    15
  extinction of, 3, 120, 303
  fossils of, 7, 19
  grazing, 17–18
  mammal, took control in Tertiary,
    15
  migrated south during early glacia-
    tion, 102

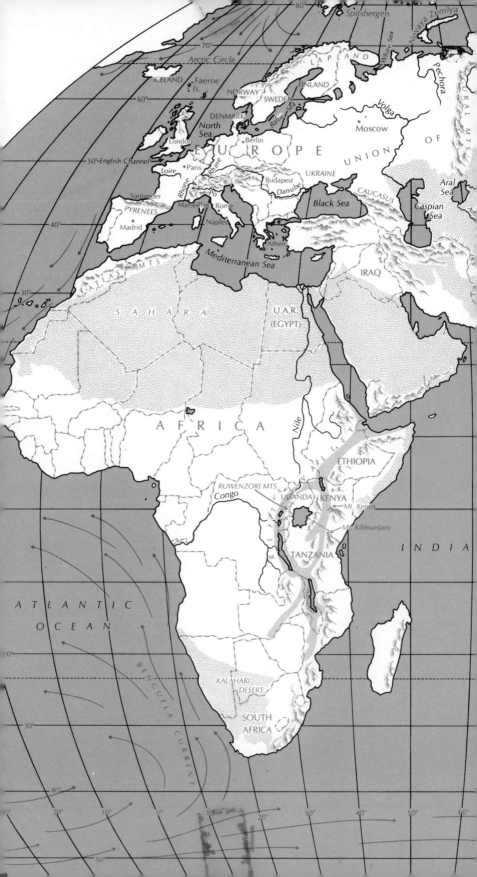